Women in the History of Philosophy and Sciences

Volume 3

As the historical records prove, women have long been creating original contributions to philosophy. We have valuable writings from female philosophers from Antiquity and the Middle Ages, and a continuous tradition from the Renaissance to today. The history of women philosophers thus stretches back as far as the history of philosophy itself. The presence as well as the absence of women philosophers throughout the course of history parallels the history of philosophy as a whole.

Edith Stein, Hannah Arendt, and Simone de Beauvoir, the most famous representatives of this tradition in the twentieth century, did not appear form nowhere. They stand, so to speak, on the shoulders of the female titans who came before them.

The series Women Philosophers and Scientists published by Springer will be of interest not only to the international philosophy community, but also for scholars in history of science and mathematics, the history of ideas, and in women's studies.

More information about this series at http://www.springer.com/series/15896

Sigridur Thorgeirsdottir · Ruth Edith Hagengruber
Editors

Methodological Reflections on Women's Contribution and Influence in the History of Philosophy

Springer

Editors
Sigridur Thorgeirsdottir
Department of Philosophy
University of Iceland
Reykjavik, Iceland

Ruth Edith Hagengruber
Department of Philosophy, Center for the History of Women Philosophers
Paderborn University
Paderborn, Germany

ISSN 2523-8760 ISSN 2523-8779 (electronic)
Women in the History of Philosophy and Sciences
ISBN 978-3-030-44420-4 ISBN 978-3-030-44421-1 (eBook)
https://doi.org/10.1007/978-3-030-44421-1

This Springer imprint is published by the registered company Springer Nature Switzerland AG
The registered company address is: Gewerbestrasse 11, 6330 Cham, Switzerland

Series Foreword

Women in the History of Philosophy and Sciences

The history of women's contributions to philosophy and the sciences dates back to the very beginnings of these disciplines. Theano, Hypatia, Du Châtelet, Agnesi, Germain, Lovelace, Stebbing, Curie, Stein are only a small selection of prominent women philosophers and scientists throughout history.

The Springer Series *Women in the History of Philosophy and Sciences* provides a platform for publishing cutting-edge scholarship on women's contributions to the sciences, to philosophy, and to interdisciplinary academic areas. We therefore include in our scope women's contributions to biology, physics, chemistry, and related sciences. The Series also encompasses the entire discipline of the history of philosophy since antiquity (including metaphysics, aesthetics, and philosophy of religion). We welcome also work about women's contributions to mathematics and to interdisciplinary areas such as philosophy of biology, philosophy of medicine, and sociology.

The research presented in this series serves to recover women's contributions and to revise our knowledge of the development of philosophical and scientific disciplines, so as to present the full scope of their theoretical and methodological traditions. Supported by an advisory board of internationally esteemed scholars, the volumes offer a comprehensive, up-to-date source of reference for this field of growing relevance. See the listing of planned volumes.

The Springer Series *Women in the History of Philosophy and Sciences* will publish monographs, handbooks, collections, anthologies, and dissertations.

Paderborn, Germany
Cleveland, USA
Vercelli, Italy

Ruth Hagengruber
Mary Ellen Waithe
Gianni Paganini
Series editors

Introduction: Methodological Reflections on Women's Contribution and Influence in the History of Philosophy

The history of women philosophers stretches back as far as the history of philosophy itself. There exists a vast amount of philosophical writings and ideas by and about women philosophers from antiquity to the present (Waithe 1987–1995). However, the history of philosophy has hitherto mostly been told as a history created by men. This still holds true not only with regard to the established idea of philosophy as it is presented in the tradition of Western philosophy. All over the world, philosophy continues to be presented more or less as a male genre. Intensive research of women philosophers from the last 30–40 years proves that the history of women philosophers exists, that it is as old as any tradition in philosophy and that it has shaped the tradition of philosophy all along (Green and Hagengruber 2015).

The questions raised in that context were initiative for reflecting on methodological issues regarding the dismissive practice in the history of philosophy around the participation of women, and the inclusion and incorporation of women thinkers into the history of philosophy. In 2015, Sigridur Thorgeirsdottir as the Jane and Aatos Erkko professor at the Helsinki Collegium for Advanced Studies, Sara Heinämaa who leads the Subjectivity, Historicity, Communality Research Network at the University of Helsinki, and Martina Reuter from the University of Jyväskylä organized a conference on *Women in the History of Philosophy: Methodological Reflections on Women's Contributions and Influence*. The questions raised gave reason for gathering philosophers to reflect on methodological questions regarding the dismissive practice in the history of philosophy around the participation of women, and the inclusion and incorporation of women thinkers into the history of philosophy.

Contemporary research into forgotten women philosophers confirms that women throughout the ages participated more actively than commonly acknowledged in the development of philosophical thought. Women thinkers were less solitary figures in philosophical communities than commonly assumed. They were thinkers who responded to their philosophical predecessors and their contemporaries. They had impact on the philosophical thinking of their times. The rewriting of the history of philosophy by telling "her story" therefore displays a more complex picture of the past than "his story" has shown up to now.

The methodological questions that the study of women in the history of philosophy raises are manifold if we understand methodology as the theoretical analysis of the methods applied in the research of women thinkers in the past. A methodology offers the theoretical foundations for understanding which methods are most suitable to this type of scholarship within the history of philosophy. Obviously such a methodology is a feminist one that has a point of departure in the view that women have been silenced, marginalized, and excluded in the writing of histories of philosophy and in canons as represented in lexica, compendiums, and academic curricula. The ideology of sexual difference that has permeated the philosophical tradition may explain the prejudiced view of women as lesser thinkers than males is not applicable to the study of women in the history of philosophy. Women's philosophies cannot be understood as the negative counterpart to male philosophies in the historical study of philosophy. Neither can women thinkers be seen as a unified group pitted against a unified group of male thinkers. There is no duality in the history of philosophy with women representing one clearly demarcated stream of thought and men another one. The past of philosophy is, like the study of women thinkers shows, a much more pluralistic history. The past of the history of men philosophers is also more pluralistic in terms of ideas about sexual difference than often assumed from modern and contemporary perspectives. There is at times a latent feminine voice in texts that also needs to be excavated, like a several of the authors in this collection point out. The study of women philosophers and their works sheds light on an all too simplified view of this past, disclosing new possibilities of understanding it, yielding a richer picture of the human being than the traditional dualistic schemes of sexual difference offer. The pariah position of women in philosophy was often also as a source of freedom for many of the rediscovered thinkers experimented with different styles of philosophizing. We encounter attempts to broaden the notion of the ideal of the philosopher and efforts to extend the idea about what counts as philosophy. Latent or overt attempts of extending the ideal of the philosopher can be seen in symbolic figures such as Diotima, the goddess in Parmenides' poem, Sophia, and Lady Philosophy, as discussed in several of the contributions in this collection.

The contributors to the volume examine women thinkers as inventors and developers of ideas and as initiators of new modes of asking philosophical questions. That generates questions on how and why the disappearance of female contributions has affected philosophy, and the ways in which their re-emergence can transform the field of academic philosophy in terms of canon, curricula, and philosophical styles. The incentive for the contributions to this volume is to reflect the methodologies, both those that have caused the exclusion of women in philosophy and those that made it possible for them to become a part of the history of philosophy.

The following contributions on methodological reflections regarding the inclusion and exclusion of women in the canon and curricula of philosophy are divided into three sections. In Part I "Methodology", the authors who have all done extensive work in this area reflect on canonical exclusion and methodologies of inclusion in the writing of the history of philosophy. In Part II with the heading

"Rewriting the History", the contributing authors reflect on the early stages of the philosophical tradition and on how this initiation was already branded by a sexual difference that determined the further historical development of philosophical culture and discourse. In Part III on "Reflecting the Content", philosophers of the twentieth century who decisively shaped the course of contemporary philosophy are examined. Two of the contributions in this collection are written from a transcultural perspective, bringing together Eastern and Western perspectives on these questions of methodology. These contributions are a further indicator for the work that is ongoing in introducing other philosophical cultures to predominantly Westernized philosophy traditions. At the same time, they shed light on commonalities in the concerns of women philosophers across cultural boarders, beyond their shared love of wisdom.

We are grateful to the authors to have been willing to contribute to an overview on the methodical reflection on the history of philosophy, on how women have influenced this history, and how and why the methodical instruments of the commonly known historiography of philosophy have to be rethought in light of this challenge. Their methodologies represent their philosophies of the diverse methodologies of analyzing women's exclusion and contribution to their respective philosophical traditions.

Finally, last but not least, we thank Catrine Val for the permission to include some of her suggestive pictures of women philosophers of the past. In her photographs. Catrine Val imagines how women philosophers and their ideas can be interpreted in art. These pictures and many more from her work on *Philosopher Female Wisdom* were exhibited at the University of Helsinki during the conference this book grew out of.

Part I Methodology

Mary Ellen Waithe's *A History of Women Philosophers* that started in 1987, providing a four-volume impressive presentation of more than seventy philosophers, from 600 BC to the twentieth century did more than anything else to open up the field of contemporary research into women in the history of philosophy. After a long period of silencing women, the reader found names, texts, and interpretations of women thinkers from all major eras of philosophy. The recovery of the works by women philosophers that has taken place since has caused a significant change in the professional perspective on philosophy's history. Waithe's opening article of this volume "Sex, Lies, and Bigotry: The Canon of Philosophy" consists in a methodological reflection on the reasons for an exclusive canon and how it has been changed with the inclusion of women philosophers. Sex, lies, and bigotry were the causes of ostracizing women and the reason for how the "true" canon of philosophy was sabotaged. Waithe distinguishes between a historical canon as the canon that has excluded women thinkers and a compendium of philosophical works which includes lost and unknown, as well as forgotten and omitted works and those works that have not "withstood the test of time". Waithe assumes that from this compendium, a true canon can emerge, if the process of emergence is not disturbed by contextual or damaging interests. Categories of race or gender have functioned as excluding categories that damaged this selective process and the canon we have is shaped by it. Waithe urges us to rethink the canon because a reweaving of the tapestry of the historical canon is needed in the interest of philosophy, to recover from a "toxic" tradition. This tradition became more toxic with the onset of the modern era, print media, and the institutionalization of sciences within modern universities. The exclusion of women philosophers and their works became more systematic as apparent in encyclopedias and histories of philosophies. Waithe concludes that we need time to rethink our history and our canon of philosophy in light of the thousands of works omitted in the historical canon.

In "Women in the History of Philosophy: Methodological Reflections on Women's Contribution and Influence" **Charlotte Witt** explores the question of how common ideas about the history of philosophy relate to the question "Why the history of philosophy matters to feminist philosophy?" Debates over methods and purposes of those philosophers who engage with the history of philosophy usually oppose the reconstructive method of the contextualists and the analytic approach of the "appropriationists", and it is questionable if this discussion is fruitful for the feminist approach. According to Charlotte Witt, there is no doubt that philosophy is a combination of both methods, as philosophy is in itself always an intrinsically historical discipline insofar as the dialectical discourse of the present consists in an engagement with the past. However, as Witt adds, with this approach the exclusion of women from the history of philosophy and the refusal to recognize feminist research cannot be grasped. The historic retrieval of women philosophers of the past is a work of *recognition* in itself that makes the feminist approach special in its own way. History demonstrates the inadequate representation of women which shows an epistemic and a political bias of philosophy. Feminist work consists in the intellectual act of seeing what has been invisible and encourages the normative act to recognize this. The new method Witt presents provides a critique of the canon, the appropriation of canonical philosophers, and a revision of the history of philosophy as we are used to it, like Witt illustrates with the example of Kant. One the one hand, we admit our indebtedness to Kant, but on the other hand, we can no longer whitewash or overlook Kant's racism and sexism.

In "'Context' and 'Fortuna' in the History of Women Philosophers: A Diachronic Perspective" **Sarah Hutton** discusses how the resonance of women in the history of philosophy is itself a result of a historical approach in philosophy. The history of their reception is their "Fortuna" that cannot be separated from their original historical context. Hutton comes to this insight by reflection on the development of the study of women philosophers of the past. The first era of the contemporary study of women philosophers was motivated by the concerns of the women's movement. There was a lack of awareness of the philosophical and historical context, resulting in readings that implanted modern ideas into earlier thinkers. Feminist interpreters highlighted themes particular to women in hopes to "reconstruct a women's tradition in philosophy", presupposing that women think differently. This approach resulted in a widespread negative view of rationality which was seen as masculine. The opposite was the case at the beginning of the contextual approach. Women philosophers were often interpreted as minor peers of the great male thinkers they were associated with. Elisabeth of Bohemia, Emilie Du Châtelet, and Anne Conway were in such historical reconstructions hardly visible for feminist discourse, but viewed in their productive contributions for philosophical discourse. Thus, the segregation of the history of women thinkers is an essential aspect of the recovery of their past contribution. The later phase of the contextual approach enables us therefore, so Hutton, to discuss women's philosophies of the past without assigning minor status them. Fortuna and context therefore both need to be considered in reworking, retracing, and re-applying the arguments and ideas of women philosophers of the past. There is necessarily a mismatch, a diachronic perspective, a gap between past and present interpretations that is integral to the study of the history of philosophy. Not only does it lead to the recovery of many figures who formerly were lost. The recovery of many thinkers who emerged from the contextual scenery was a result of a "mismatch" of historical context and present day philosophical interest. This requires us to ask, why we read today, as we read today and inspires us to reflect on the transformation we undergo in our view of the past and present of philosophy.

In "The Stolen History—Retrieving the History of Women Philosophers and its Methodical Implications" **Ruth Edith Hagengruber** insists that the history of women philosophers is as long as the history of philosophy. The forgetting and excluding of women in the tradition of philosophy deprived women of constructing their own identity. Re-reading and re-evaluating history is much more than a contextual assessment of the relevant conditions of a specific time period. Doing history is not collecting contextual narratives from facts to serve the interpretation of the present dominant agenda . Doing history as a methodical and methodological approach to philosophy is a unique and indispensable means to widen contemporary insights by becoming aware of facts. The history of philosophy is no quarry of ideas from where you extract the narratives that provide the concepts you are interested in. This narrow-minded approach is a methodical misuse of the power of history. Much more than a quarry where you obtain the material you asked for and consolidate the opinion you have already formed, the reading and re-reading of history can be an instrument for change, for thinking anew, for doubting convictions and for gaining new insights. Everything history delivers may turn out differently from what is expected. And the rarer the ideas one is able to embrace, the more revealing their reception within the history of ideas can become. The mind widens, so Hagengruber concludes, as it develops the ability to embrace different points of view.

In light of the above, research into the history of women philosophers brings to the fore new facts of a rich tradition. Including the findings of this research in the book pile that contains the history of philosophy, provides a kingdom of new ideas to look for insights that offer alternatives to conventional ones.

Part II Rewriting the History

As a scholar of ancient Greek philosophy **Vigdis Songe-Møller** has written widely about notions of femininity and how women are viewed as superfluous in ancient philosophical thought. Women are quite present as characters in tragedies and comedies, but not in philosophical literature. There are however two important exceptions. In Parmenides's poem, a goddess has the central role of speaking the truth to the young philosopher, and in the homoerotic gathering of the *Symposium* of Plato, Diotima has as a priestess the role of indicating truth as the goal of Plato's philosophy. In both cases, these female figures represent the other who educates the young man in the poem and teaches Socrates. Songe-Møller displays how the feminine other is represented by the goddess as difference, birth, and death. The goddess nevertheless teaches the philosopher the denial of origin and change for the sake of timeless presence of true being. Powerful thinking is defined as transcending the needs of the human world which seems to contradict the female character. Parmenides, who narrates the journey of the philosopher, confronts him with the metaphors of light, unveiling and birthing, symbolized in passing by the vaginal cleft. When the young man has arrived at the imperium of the goddess, he has to listen to her commands, to preserve and pass her story. Here he learns that being and thinking are the same, mutually dependent and mutually constitutive. So "ironically" the goddess teaches the young man a truth which expels the female. In Plato's *Symposium*, Songe-Møller observes a dialectical change that goes in an opposite direction to the poem of Parmenides. Diotima's speech presents an active part instead of the passive love of beauty. In her story, the traditional roles of the sexes, presented in Penia and Poros, resource and need, become inverted. Penia is active and even rapes the passive Poros. Songe-Møller calls it a "radical feminization of Eros and thus of philosophy" as Eros' activity is said to be the one that belongs exclusively to women, namely giving birth. Reproduction is what mortals have in place of immortality. Songe-Møller hence concludes that in the poem of Parmenides, the goddess is used to expel the feminine, and in the *Symposium*, Diotima is used to include the female in Plato's philosophy. Her interpretation is hence an example of a methodology of uncovering the uses of ideas associated with women and the feminine. She uncovers the feminine in these texts, and displays how it is undermined by the male order in these ancient Greek texts.

Robin R. Wang has been influential in introducing ideas about sexual difference in the history of Chinese philosophy, both in the Confucian and in the Daoist traditions. In "A Journey of Transformative Living: A Female Daoist Reflection" Wang illuminates the primary role of the feminine and the maternal in Daoist thought, originating from Laozi's *Deadening* or *Classics of Way and Its Power* (that possibly dates from as recently as the third century B.C.E.). She also argues for its necessity as an intervention into sexism and gender inequality in Chinese contemporary culture. This may at first sight strike as odd given the fact that Daoist and

Confusion traditions have throughout the centuries justified and legitimized patriarchy. By disclosing the potential of Daoism for feminism and equality concerns, Wang goes against mainstream patriarchal interpretations of these interpretations and uncovers a more pluralistic philosophical past as a resource for feminist thought and for the sexes to rethink their gender identities. Wang's interpretation also offers an interesting intercultural perspective. The gendered world of Daoism is different from Aristotle's male-female cosmos which is characterized by a strict hierarchical order of the active masculine and the passive feminine. There are no male images such as father/son or force, strength, and aggression that are associated with Dao. Daoism however makes the philosophical imaginary of the feminine a privileged locus of an embodied, experiential way of thinking and knowing. The rhythms of nature inherent in the feminine are important, the soft and the yielding that Dao represents, for making women and men aware of their relation to femininity, and to help them overcome the denial of maternal origin that is basic to any patriarchal thinking.

With her interpretation of Lady Philosophy in the early medieval text of Boethius, *The Consolation of Philosophy*, **Sigridur Thorgeirsdottir** offers a novel reading of Philosophy as a feminine personification in this text which was one of the most widely read texts of the middle ages. Here a symbolic womanly figure is interpreted as representing a philosophy that accounts for sexual difference. The title of "The torn Robe of Philosophy" refers to how Lady Philosophy describes how fake philosophers have torn her robe and adorned themselves with pieces of it as if it were their own philosophy. This metaphor of the torn robe can be seen as a metaphor for how philosophies of women have been treated in the history of philosophy. It can also be seen as a metaphor for the interpretations of Philosophy as a woman in the history of the reception of Boethius' text. Lady Philosophy's historical background in the goddess Sophia is repressed. It is necessary, so Thorgeirsdottir, to interpret this figure as a tension between remnants of pagan elements and Christian-Platonic views. The history of interpretations of Boethius's text displays efforts to neutralize and degenderize the figure of Philosophy. This fact testifies to the no-place of women and the feminine in the history of philosophy that Luce Irigaray has discussed in her readings of classical texts of the Western philosophical canon. In her interpretation, Thorgeirsdottir discusses the figure of Philosophy in terms of embodiment, the feminine and maternal origin. The imprisoned Boethius who awaits his death sentence in confinement has a dialog with Philosophy in which she sings for him, recites, and ponders on the question of life, death, and fortune. She works with feelings in thinking and thinking in feelings to help Boethius figure out the big questions of life and death to come to terms with his grim destiny. Thorgeirsdottir argues that Philosophy represents ancient meanings of the noun *sophia* that include practical, embodied, and sensual knowledge and not mere theoretical wisdom.

Part III Reflecting the Content

In "Reconsidering Beauvoir's Hegelianism" **Karen Green** discusses how Hegelian philosophies of history have permeated histories of philosophy as histories of a masculine spirit that silence women's contribution to the development of philosophical thought. One of the reasons why the history of women in philosophy is a relatively new field may have to do with the fact that Simone de Beauvoir's philosophy of sexual difference is also determined by the Hegelian model according to which the womanly is identified with the other and the objectified according to Hegel's master-slave dialectics of historical progress. Beauvoir did acknowledge how thinkers like Christine de Pizan defended their sex against the misogynistic tradition. Beauvoir's Hegelianism remains nevertheless intact in terms of her view of the misogynistic history, and she does not, so Green, encourage research into the history of women in philosophy.

According to Green, there are two options for studying the history of women philosophers. One of them is to study women thinkers that have been closely affiliated with famous male philosophers "which requires immersion in the male stream from which they run as minor tributaries". The other option is an alternative history of women's ideas, which like women's studies, most men choose to ignore. Green therefore comes to the conclusion that what is needed is a "cultural double helix, a sophisticated history in which we recognize both the evolution and development of men's ideas, and the evolution and development of women's ideas, as well as the complex interaction between them." For Green, this task is of the utmost importance for philosophy because the solidarity of women is impossible without a history of philosophy that omits its women. The remembering of future women philosophers also depends on recovering this past.

With Hegel as the last great thinker of a philosophical system, **Tove Pettersen's** interpretation of Beauvoir's philosophy as a rejection of philosophical systems comes as a testimony to the diverse ways in which Beauvoir reacts to Hegel's philosophy. In Simone de Beauvoir and the "Lunacy Known as 'Philosophical System'" Pettersen shows how Beauvoir distanced herself from philosophy by denying to call herself a philosopher (which delayed that she was taken seriously as a philosopher in her own right). She has this in common with many great thinkers, be it Nietzsche, Wittgenstein, Heidegger, or Arendt who were critical of institutionalized and some canonical forms of philosophical thinking. What was of utmost importance to Beauvoir, as Pettersen makes clear, was that she felt she could not be true to her own experience and her feminine condition in her philosophizing given the parameters of traditional philosophical methods, concepts, and definitions. The traditional voice of philosophy is a masculine one. These struggles that Pettersen traces back to her early *Diary* helped her to open up a new field: feminist philosophy. Her commitment to cultivating one's different voice is beautifully illustrated in her remark that the woman "who chooses to reason, to express herself using masculine techniques, will do her best to stifle an originality she distrusts. … She may become an excellent theoretician and a reliable scholar; but she will make herself repudiate everything in her that is 'different'." The point of philosophy, of the past and of the future, is to explain the mystery of the "universe and of my own existence" like the young Beauvoir writes. So Beauvoir speaks against the traditional voice of philosophy as a masculine voice that presents itself as representing a neutral stand and speak on behalf of mankind. Her perceptive view of philosophy, she claims, stems from her lack of philosophical originality. She views herself as a literary person because it allows her to express her lived experiences. This self-defamatory view is based on her critical view of philosophy as abstract system building, out of touch with experience that she claims to be nothing less than a form of lunacy. So Pettersen comes to the conclusion that Beauvoir did not want to be termed a philosopher based on how philosophy was measured by the traditional standards. Reading Beauvoir's oeuvre in this way, like Pettersen does, shows how women thinkers have reshaped and reformulated the domain of philosophy by transgressing accepted methods and styles. One of the tasks of the study of women in philosophy is hence to show how our philosophical past is broader and more pluralistic than hitherto acknowledged, including an oeuvre like Beauvoir's that includes philosophical, literary, and essayistic texts. Overseeing this plurality amounts to a testimonial injustice inherent in canonizations and histories of philosophy.

Along with Beauvoir, Hannah Arendt is certainly among the most significant thinkers of the twentieth century. Like Beauvoir, Arendt maintains a distance to philosophy with her claim that she does political theory. In her paper on "Arendt, Natality, and the Refugee Crisis" **Robin May Schott** also discusses a thinker who has a critical view of the feminist movement of her time. The reason was that she thought that the feminist movement was occupied with small issues and disregarded larger political systemic injustices. Feminist interpreters have nevertheless not been unified in their view of Arendt's philosophy which has both been interpreted as phallocentric and as gynocentric. Having left these debates behind, Arendt's philosophy emerges as an important toolbox for feminist analysis of diverse political phenomena. Arendt's concept of natality makes in Schott's view a revolution in philosophical thought, with its inherited focus on mortality as the horizon on which meaning takes place. Schott employs the concept of natality to reflect methodologically and normatively on the current refugee crisis. Like Pettersen's interpretation of Beauvoir, Schott shows how Arendt's philosophy is transformative for philosophy in general by taking the feminine experience of birth (although she does not discuss actual physical birth) and to be born to bring a new perspective on life, birth, and death to a mortality-centered philosophical tradition that ranges from Socrates to Heidegger. Schott's application of the concept of natality sheds light on the limits of the influential philosophical concepts of biopower and biopolitics for analyzing phenomena like the refugee crisis. This comparison is also interesting from the perspective of questions of methodology because it reveals both the disregard and the belated acknowledgement of the impact of women thinkers like Arendt. Arendt's concept of natality as a concept to analyze critically actual humanitarian crises and their political backgrounds illustrates moreover a change that has taken place within Western philosophy over the last half a century or so.

Arendt's rejection of belonging to what she analyzed as an apolitical approach of a philosophical tradition as well as her criticism of the idea that philosophers speak in a universal and neutral voice have become accepted even though Arendt as a motor of this change has not been adequately acknowledged.

In her text on "The Feminine Voice in Philosophy" **Naoko Saito** follows a similar path of undermining the philosophical idea of a detached, neutral, impersonal subject of philosophy. Like Luce Irigaray sees a great potential for growth and diversification in philosophical knowledge in an encounter between the sexes, Naoko Saito advocates an alternative mode of philosophical thinking that consists in a new conversation between the sexes. She bases this on her idea of philosophy as translation that implies that any meaning comes to be through the act of a translating in the very general sense of translating a sense into word, an idea into interpretation like a word into another language. The idea of translation destabilizes the philosophical subject as an autonomous agent and unsettles predominant ways of thinking. In her interpretation of Cavell's concepts of the father tongue (writing) and the mother tongue (speech), Saito develops further his ideas about recovering the feminine voice in philosophy that has been suppressed. She does not understand the feminine in any essentialist sense but rather as voice and language that destabilizes predominant ways of thinking that harden the dichotomies of reason and emotion, of justice and care, and of male and female. From the point of view of methodological reflections, the conversation Saito introduces needs to take place within philosophy in a similar manner she reads male thinkers like Cavell and Emerson, by making the feminine voice in their texts better explicit. Her own interpretation of these texts is hence a form of conversation that illustrates her point that the conversation is more than a sum of its parts because it generates something new. Her approach has implications for transgressing other boarders and divides. Occidentalism and Orientalism also need to be freed out of an oppositional structure that discloses a space for something new to emerge. The virtues of such an approach are those of listening, responsiveness to difference, and a willingness for change.

Saito's focus on a sexual binary should in this context not be understood as a return and rehabilitation of a binary culture of two sexes. She is more interested in the oppositional and dualistic structures that have permeated philosophical thought. These dualities have for the most part been associated with a hierarchical duality of sexual difference. For that reason, the history of philosophy leaves us with a legacy that is permeated by conceptual pairs, schemes, and structures of oppositions that are based in ancient ideas about a sexual difference. These schemes need to be rethought.

In "Iris Murdoch on Pure Consciousness and Morality" **Nora Hämäläinen** takes issue with a recurrent theme of the twentieth century women thinkers discussed by the contributors to this collection. It is fitting that the last contribution in the collection offers perhaps the most radical idea of a philosophy that attempts to extend or transgress the given parameters of a philosophical tradition that concentrates on disinterested pursuit of knowledge, and the exercise of thought for its own sake. Hämäläinen interprets Murdoch's philosophy of pure consciousness and morality as an interesting form of bridge-building between traditions and schools of philosophy. With her philosophy, Murdoch opens up a space between stifling divisions where new possibilities for philosophy emerge. This methodology illustrates quite well how a certain outsider position can be productive for philosophy. With her fresh approach Murdoch builds bridges between theoretical and practical philosophy, between impersonal abstract thinking and efforts to become a better person through philosophical reflection. Murdoch goes beyond the virtue ethics of Elizabeth Anscombe, Mary Midgley, Philippa Foot, and their analytic contemporaries by elaborating her own kind of moral phenomenology of pure consciousness. She does not focus on consciousness as a foundation of knowledge in the tradition from Descartes to Husserl. Her interest lies in reflecting the experience of moral life. It is not enough to ponder on philosophical arguments. Arguments are means to understanding but not the substrate of the type of philosophical thinking that she is interested in understanding. Her accomplishment lies in conveying an understanding of philosophy as a fairly technical pursuit and as a work on oneself, to understand oneself and one's desires, longings, and circumstances better. In that sense, Murdoch combines the personal and the impersonal aspects of philosophy in a novel fashion. Her philosophy relies in this regard on our innate ability for non-intellectual perceptiveness and natural virtue that we can connect with and can be a source for more abstract reflections. Murdoch hence produces something that Hämäläinen describes as a philosophy of intelligent wisdom, offering a bridge between contemporary philosophies of wisdom as a pursuit of a good life, and more abstract, theoretical philosophy. This attempt may be said to unify most of the women thinkers presented and discussed in this collection. Their common denominator is a sense for philosophy as borne out of lived experience that abstract philosophical reflection requires for getting closer to life as we live it.

Paderborn, Germany — Ruth Edith Hagengruber
Reykjavik, Iceland — Sigridur Thorgeirsdottir

References

de Beauvoir, S. (2010). *The second sex*. (trans. Borde C., & Malovany-Chevallier S.). New York: Knopf.

Green, K., & Hagengruber, R. (Eds.). (2015). The history of women's ideas. *The Monist, 98* Nr. 1. OUP.

Waithe, M. E. (1987–1994). *A history of women philosophers* (Vol. 4). Dordrecht, Boston, Lancaster: Martinus Nijhoff Publishers.

Contents

Part I
Methodology

Chapter 1
Sex, Lies, and Bigotry: The Canon of Philosophy

Mary Ellen Waithe

Abstract In "Sex, Lies, and Bigotry: The Canon of Philosophy" I explore several questions: What does it mean for our understanding of the history of philosophy that women philosophers have been left out and are now being retrieved? What kind of a methodology of the history of philosophy does the recovery of women philosophers imply? Whether and how excluded women philosophers have been included in philosophy? Whether and how feminist philosophy and the history of women philosophers are related? I also explore the questions "Are there any themes or arguments that are common to many women philosophers?" and "Does inclusion of women in the canon require a reconfiguration of philosophical inquiry?" I argue that it is either ineptness or simple bigotry that led most historians of philosophy to intentionally omit women's contributions from their histories and that such failure replicated itself in the university curricula of recent centuries and can be remedied by suspending for the next two centuries the teaching of men's contributions to the discipline and teaching works by women only. As an alternative to this drastic and undoubtedly unpopular solution, I propose expanding the length and number of courses in the philosophy curriculum to include discussion of women's contributions.

It has been three decades since the appearance of Volume 1 of *A History of Women Philosophers,* part of a series that some have credited with causing western philosophers to question the accuracy of the Canon of Philosophy that forms the framework for higher education in that discipline. In 30 years a lot has changed in the profession, but hardly enough. The questions explored at the 2015 Helsinki conference merit serious consideration not only by those of us with research programs related to women and philosophy but by the entire profession. Accordingly, I will turn to questions addressed at that conference and share my thoughts.

M. E. Waithe (✉)
Department of Philosophy and Comparative Religion, Cleveland State University, Cleveland, OH 44115, USA
e-mail: professorwaithe@sbcglobal.net

S. Thorgeirsdottir and R. E. Hagengruber (eds.), *Methodological Reflections on Women's Contribution and Influence in the History of Philosophy*, Women in the History of Philosophy and Sciences 3, https://doi.org/10.1007/978-3-030-44421-1_1

1.1 What Does It Mean for Our Understanding of the History of Philosophy that Women Philosophers Have Been Left Out and Are Now Being Retrieved?

In this section, I will examine interrelated issues: (a) whether recoveries of works by women philosophers in recent decades amount to a significant change in the historical canon; (b) whether we can have confidence in the historical canon; and, (c) whether that canon has caused harms to the discipline and profession of philosophy.

A. Degree of change

Some who teach and write histories of philosophy may be confident that they know that history, its major contributors from Pythagoras to Poincaré, and its minor contributors from John of Tynnemouth (*huh*?) to Henry of Brabant, i.e., Philosophy's historical canon (HC). The HC is drawn from what I refer to as the Compendium: all works of philosophy, whether or not they are presently known. The Compendium (C) includes the HC and in addition, works that are lost but whose titles are remembered in our histories, works that are completely unknown but that are philosophical, works that have been forgotten or omitted from our histories, and recent works that have not yet withstood the test of time. By definition then, historians of philosophy cannot know the entire C, or even most of it. But they are well versed in that subsection of C that is the HC: significant works, insights, arguments and their authors, important schools, movements, milestones, and the comparatively minor players whose contributions sharpened the debates or provided historical continuity to movements.

Under normal circumstances, the informal HC evolves gradually from the C in an almost-organic way: through an informal process rather than by a formal methodology. Recent works of philosophy (by definition, part of C) generate interest among contemporary philosophers who study those writings. Those writings—their arguments, insights, etc.—get translated, discussed, cited, etc. with such frequency that they become the subject of journal articles, colloquia, symposia, etc., and in time are introduced into philosophy curricula of many institutions. There is no formal procedure, no minimum number of citations or colloquia needed, no votes by members of the profession required for a work (and its author) to become included in that subsection of C that is the HC. In less than a century following its publication, a work of philosophy can become part of our discipline's historical canon.

So long as most meritorious works of philosophy make it into the HC through the same organic fashion, the HC can reliably be considered to be the true canon of philosophy (TC). The TC will, therefore, be a more inclusive HC than it would be were the HC to exclude works and authors on the basis of gender or race. If a work from C is preselected for omission from HC on spurious grounds such as the race or gender of its author, it is as though only a single gardener were to use inorganic pesticides in her yard: the effect on the ecosystem will be negative, but slight. However, if the relatively organic process of growing the HC were abandoned and spurious preselection was to be used for many centuries in succession, then, like any natural system in which many gardeners used inorganic pesticides, the effect on the environment that is the discipline of philosophy will be toxic. And indeed it now is.

Under such circumstances, the HC cannot be considered to be the TC. Historians of philosophy–those who know the HC—have only an incomplete, confused, vague notion about what the TC is. They are in the same sorry, but the exciting circumstance that historians of religion were in when the Dead Sea Scrolls were discovered.[1]

Discovery of the Dead Sea Scrolls heralded a degree of change in the Judaic and Early Christian Canons. Prior to their discovery, Judaism was believed to have been known: a rich tapestry, albeit somewhat tattered, with some threadbare sections, and ragged edges. But it seemed sufficiently intact that we could tell what its basic tenets and history was, what figures, works, and teachings were canonical. With the discovery of the Dead Sea Scrolls, no one could claim to know either the history or the Canon of Judaism or early Christianity. Much will need to occur before that claim can validly be made. Translations and analyses of the scrolls by literally hundreds of experts are needed.[2] Interpretations and implications need to be tossed about at professional conferences and in the professional press: argued for, discussed, and rethought. Meanings of strange words[3] from ancient dialects needed to sink in and their contextual clues investigated. Competing opinions needed to be considered, facts needed to be investigated and reconsidered in light of the previously-known histories of Judaism and early Christianity. Identities of historical persons and their relationships needed to be rethought.[4] A new tapestry was being woven from many threads of Judaism's HC. It was not immediately knowable whether a markedly different picture of its HC would emerge, or whether a picture that was more or less the same but with subtler details would result.

---

[1] I owe this insight to Ms. Nancy Tomaselli, a member of the Society for the Study of Women Philosophers.

[2] With respect to the history of Ancient Philosophy and its Canon, for example, the previously untranslated fragment by Aesara of Lucania from her book *On Human Nature* has now been translated (Waithe 1987). Its relation to the Platonic *corpus* needs to be discussed: was she adopting a Platonic concept of the tripartite human soul? Had he adopted her analysis? Or, were they both explicating views that were generally known and generally believed to be true? Scholars hold different views of the relative antiquity of her Greek dialect, but much more discussion and analysis is needed before we can with assurance credit that view of the soul to Plato alone.

[3] In the history of women's contributions to Philosophy, an example of such an expression is '"Εκδοεως παραναγνωσθεώης τή φιλοσόφ,ῳ θυγατρί μου Ὑπατίᾳ' appearing in Theon of Alexandria's footnote where he acknowledges Hypatia's responsibility for the *Commentary on Book III* of Ptolemy's *Syntaxis Mathematica,* 'On the Motions of the Sun' (Rome 1943, p. 807). Elsewhere (Waithe, *HWP 1*) I have suggested that Copernicus likely read Hypatia's own footnote to the effect that 'for Ptolemy to be correct, the sun would need to be in two places and that is impossible,' and thus started Copernicus rethinking the Ptolemaic geocentric universe.

[4] With respect to the history of women's contributions to philosophy, the work *On the Harmony of Women* by Perictione I (Waithe, *HWP1,* 32) raises all sorts of questions: Is this the same Perictione as Plato's mother? If she was a philosopher, what was her influence on her son's philosophical development?

Like historians of religion, historians of philosophy now must re-examine the tapestry that is the history of philosophy and its canon.[5] We do not know precisely what the history of philosophy is, but we do know that it is not what we thought it was. In the past three decades, we have discovered thousands of works by hundreds of philosophers. So many parts of C were omitted from the HC that it is simply stupid to think that a different picture of the canon won't result once these many loose threads are rewoven into the HC tapestry. The question is: what will be the degree of change?

B. A question of confidence

In the social sciences, confidence in a database, in a Compendium, that samples nearly all the available data is confidence well-placed and has high predictive value. In contrast, confidence based on sampling data that *has been preselected by race and gender* is confidence misplaced, logic defied and bigotry glorified. Why? Because—as is the case in philosophy—the omission of works by women (and non-whites, but for purposes of the present discussion I limit myself to the example of women philosophers) from the HC did not follow a chain of events similar to that which likely led first to the concealment and then, the ultimate recovery of the Dead Sea Scrolls. Works by women were not stashed in a labyrinth of caves because they were viewed to be such important works that there was a need to protect them and preserve them from marauders, from an inhospitable climate or from the passage of time. Works of philosophy by women were not hidden away; they were always part of C. So why were they viewed to have been "lost" and why do we now consider many to have been "recovered?"

Women philosophers' writings, part of the Compendium of philosophy, have for the most part been carefully preserved in multiple copies, by successive generations of scholars and librarians, and have been known by (competent) historians of philosophy from antiquity until the eighteenth century. Many of the works by pre-seventeenth century women philosophers survived the censorship of various Inquisitions only due to the prevalence of humility formulas. Such formulas usually occurred in the first pages of their writings and denied that the author claimed any authoritativeness with respect to the subject of her work. A lovely example is that of Julian of Norwich who states in the "Short Version" of her work that "I am a woman, lewd, feeble and frail…" with nothing important to say, and then in the "Long Version" continues for hundreds of pages to develop a metaphysics and epistemology of religion incorporating her view of "Christ, our Mother" (Julian, *Showings*, p. 6.). From the eighteenth until the twenty-first century, pre-eighteenth century source materials were still referred to by historians of philosophy. But, in creating their own histories and encyclopedias, these historians systematically omitted mention of women philosophers they had

[5] I am not suggesting that the thousands of works of philosophy that have been omitted from the HC are to the discipline of Philosophy what the Dead Sea Scrolls are to the discipline of Religion. It is too early to know how significant they are, yet, the point is the same: we must suspend judgment and take the time needed to rethink our history and therefore our Canon.

learned of.[6] This led philosophers who relied upon contemporary histories and encyclopedias to believe that works by women had been lost from the Compendium when in fact, out of sight, they were out of mind. The result is that contemporary professional philosophy in the west is informed and shaped by nearly all-white, nearly all-male histories, encyclopedias and other major source materials of recent centuries. (There are exceptions.) Those post-seventeenth century resources are considered to identify the canonical works of our discipline, the HC.

From the perspective of the discipline of history, the HC creates a false picture of what occurred in the past. From the perspective of the discipline of Philosophy the HC, like *Alice Through the Looking Glass*, is distorted, confused, illogical, false, misleading. In a discipline in which precision, clarity, and accuracy are highly valued, the HC yields instead vague imagery of reason as the domain of white males. We ought not to place any confidence in encyclopedias, histories, epitomies, and anthologies that are products of this distortion. Due to the scant representation of philosophical writings by women in the present HC we ought to suspend judgment as to what is the true canon (TC) of our discipline. TC is an ideal that we have good reason to believe we have not come close to reaching.

C. Many Harms/Many Fouls

There are additional senses in which we do not understand the history of philosophy because women philosophers have been left out and now are being retrieved. The legacy of writings by, and therefore the historical reputations of women philosophers have been harmed by the HC. Great philosophers—women who led entire schools of philosophy—have suffered posthumous harm through the neglect of historians of philosophy. In antiquity alone, we can mention Arete of Cyrene, Asclepigenia of Athens and Hypatia of Alexandria (Waithe, *HWP 1*, pp. 165–205). They were not "merely" female philosophers who worked with the males of their epoch whose reputations have survived; they themselves had reputations that drew students from afar. They advanced entire movements and had teachings and writings that were preserved by their student successors. This is true also for women of every later era until the early modern period. The harm done by the omission of our foremothers is, to use sports parlance, a foul. It is a foul in the sense that it violates the rule that histories ought to record the truth; it merits penalties against offenders whose reputations as historians deserve our criticism. But worst of all, the omission of women from the HC has inexorably altered the game.

The HC has inexorably altered the profession of philosophy and every subject division of it. No one really knows the histories of those divisions. If we are to know the history, and therefore the canon of any division of philosophy, we need a clear accounting of the Compendium: the works and authors whose insights and critiques brought theory in that division to its present state. This last claim is true of divisions as ancient as ethics and philosophy of mathematics as it is of divisions as relatively recent as artificial intelligence and feminist philosophy. No one really knows the

---

[6]I have in mind multivolume series such as that by Frederick Copleston and the *Encyclopedia of Philosophy* by Paul Edwards.

histories of those divisions because their source materials are mostly derived from the HC with its inorganic preselection from the Compendium for philosophers' race and gender.

1.2 What Kind of a Methodology of the History of Philosophy Does the Retrieval of Women Philosophers Imply?

By what method do we get from HC to something closer to TC? First, we need to understand how and why women's works were excluded in order to avoid repeating those errors. Second, we need to rediscover, reclaim, translate, discuss, and evaluate those works in their own right and also in the context of their relationship to works by other philosophers.

A. The law of the excluded middle

Why did women's writings appear in early histories of philosophy, get excluded for nearly four centuries, and then get reintroduced? What happened in the middle of the history of our discipline to cause the omission of philosophy by women? Was it gross incompetence and ineptitude on the part of the historians, or was it their pervasive, flagrant bigotry against women? Male historians of philosophy in the past four hundred years have left us an HC in which no contemporary philosopher can have confidence: a portion of the Compendium preselected for gender. Contemporary source materials are derived from the previous HC, updated, one hopes, by recent important writings and their authors. Newer source materials and educational programs of the discipline were mostly based upon that HC, perpetuating the preselection for gender even if entries of the most recent contributions to the discipline did not completely preselect for it. In the early twenty-first century we have an HC that is generally segregated according to gender but with token newbies added on top. Karen Warren referred to this practice as "add women and stir."

The sheer number of women philosophers (nearly two hundred as of this writing) and works (about one thousand) located by the recovery and restoration projects of the last three decades implies the need for a different methodology than the inorganically toxic process that created the present HC. In order to improve upon the HC, we need to understand the conditions and methodologies through which it was created and make corrections of those conditions and those methods.

In earlier centuries, isolated copies women's works of philosophy were preserved in hand-copied recensions, memorialized in *scholia* by those who studied and reproduced their writings. But writing is not publishing. In later centuries marked by the end of hand-copied recensions and by the beginning of mass production, it was possible to bury our early history the way the Dead Sea Scrolls were buried: waiting for our generation to recover them. The historians of the seventeenth through twentieth centuries worked in a period that followed the decline of monastic and clerical institutions of higher education. This same period is contemporaneous with the rise of secular universities, the proliferation of great private secular and public libraries, the popularization of mobile printing presses with movable type, a rapid increase in the number of commercial and academic publishers, the exploration and colonization by Europeans of nearly all of the extra-European globe, the unification of European kingdoms and principalities as civil states, the emergence of democratic movements and the decline of Papal authority.

Paradoxically, it is with the invention of the printing press, it is with the development of democracies, it is with the relative pluralization of societies in Europe that the exclusion of women from our histories and reference works began. Why? Much like the contemporary urge to have one's own blog and website, historians and philosophers of this period wanted to be published. Most publishing was of the vanity press model: writers raised money by having sponsors who underwrote the printing and binding costs and a press would publish the work. Authors had very small stock delivered to (usually only one of any major city's) booksellers. Moneyed people purchased the works. New authors of modest means who lacked sponsors typically had funds only to publish small octavos. Due to the personal responsibility for the costs of printing and binding, material from the author's source works had to be edited out for reasons of space.

The *querelle des femmes* that had raged since Christine de Pizan again heated up as the advancement of democratic forms of government threatened men's dominance in the public sphere. The misogyny of male historians of philosophy revealed itself on the editing-room floor: mention of women and their works were cut or were simply omitted from histories of philosophy even as "famous ladies" works, popular since Boccaccio (1374), enjoyed a resurgence.[7] Women, and their works, were reduced to cute little novelties of the "famous ladies" genre of books as other genres—histories of philosophy—omitted them from indices, from tables of contents, and from the content material itself. It is in the "famous ladies" books that we sometimes find mention of women's contributions to philosophy. Accomplished women were presented as novelties through a perspective that increasingly trivialized and romanticized them. For example, Émilie Du Châtelet is commemorated for rouging her nipples and for being Voltaire's lover. The fact that her particular analysis that *force vive* or energy equals the mass of object times its velocity squared (Du Châtelet, *Institutions de Physique* 1744)[8] brought physics to the doorstep of Einstein's $E = MC^2$ (energy equals the mass of object times the speed of light squared) goes undiscussed (Wade, *inter alia*).

[7]For example, William Alexander's *The History of Women from the Earliest Antiquity.*

[8]I rely here upon the excellent translation in Judith Zinsser, Émilie Du Châtelet: *Selected Philosophical and Scientific Writings*, pp. 191–192 referring to Du Châtelet, *Foundations of Physics*, section 577, where Du Châtelet discusses *forces vive* which we might (roughly) translate as 'energy': 'thus, the force of body A, which had 2 of speed and 1 of mass, was 4, that is to say, as the square of this speed multiplied by its mass.' Thus, Du Châtelet's 'the force of body A' may be abbreviated as 'the energy of A,' or as 'E'; '…the square of this speed…' may be abbreviated as 'velocity squared' or 'V^2'; and "multiplied by its mass' may be abbreviated as 'times M' to yield '$E = V^2M$' or equivalently: '$E = MV^2$'.

In this early modern period, women of means might publish poetry, pious works, etc., but generally, it was not socially acceptable for them to publish works of theology, philosophy, (canon) law, medicine or science. Various Inquisitions, state censors, and other political threats certainly had a chilling effect on the publication of philosophical works by men and women, and there were, of course, exceptions to these general limitations. But if the book was on any subject purporting to be factual, most women authors appear to have published anonymous first editions. The invention of the printing press and the various movements by civil and ecclesiastical authorities to quell unorthodox views appears to have been accompanied by the silencing of more women than men philosophers. Although fewer women than men published philosophy during this period, no women, to my knowledge wrote comprehensive histories of philosophy during this time. (Many did mention as inspirations, earlier women philosophers, but brief mention is hardly the same as either a full analysis or as a reproduction of an early text.) I know of no reason for women of this period to have written philosophy, but not to have written histories of philosophy. Between the seventeenth and twentieth centuries, it largely fell to males to record and report the history of philosophy through to the late twentieth century, and the methodologies they employed omitted our contributions. They used what I call "the lazy boy methodology." To recover the works they omitted, I recommend using "the female detective methodology."

(1) *The Lazy Boy Methodology*: Most historians of the seventeenth through twentieth centuries generally have employed what I call "the lazy boy methodology." These historians have engaged in scant primary research themselves, instead, it appears that they have reclined in their leather chairs, pipes in hand[9] ensconced in magnificent state and private libraries and copied, translated, combined and edited the source materials published by those of their predecessors whose works they could get easily their hands on. But these incomplete, inaccurate and therefore incompetent successor histories sometimes contained some clues that the lazy boy method does not work and clues as to why it miserably failed our discipline.

(2) *The Female Detective Methodology*: What does work is a method I call the "female detective method," a method I unwittingly learned in childhood by reading Nancy Drew and Agatha Christie novels. It is a philosophical method: ask the right questions, question the veracity of the answers you receive, dig further and the truth will out. I have relied upon the HC—the source materials

[9]I have encountered no women authors of HC source materials prior to the recovery movement that began in the 1980s. In the 20th century Ariel Durant with her husband Will Durant wrote for a general audience their multivolume S*tory of Civilization* without mention of women philosophers. Elizabeth Flower wrote (with Murray G. Murphey) *History of Philosophy in America*. Despite Flower's professional training as a philosopher they omit mention of women philosophers. Mary Ritter Beard wrote *Women as Force in History* (1942) and included mention of several women who were philosophers. Beard contains sufficient resources to have been a jumping-off point for my own research. Other women philosophers of the twentieth-century wrote works that were called 'histories' of particular eras or subdivisions of philosophy, but they too relied upon a small selection of early modern histories.

I've just impugned—to provide clues to the names of and works by women that would bring the HC closer to being TC. I started by searching through indices of histories and encyclopedias. When I found no mention of women philosophers I questioned whether that meant that the work mentioned none. My inner female detective told me "of course not." Case in point: Copleston's multivolume history of philosophy does not list Hypatia in the indices of its early editions. But if you turn page after page you do find a half-sentence mention of her in the section of the Alexandrian School, and the identification as her student, Bishop Synesius. (Copleston, Vol. 1.) In a later volume, Copleston mentions Kristina of Sweden in the context of her hire of Descartes. Other sources yielded that she established a philosophical academy in Rome. Like Nancy Drew and Agatha Christie, we must tie together these loose threads. So I paged through history after history, encyclopedia after encyclopedia, epitome after epitome, anthology after anthology in major university and national libraries. They eventually yield a newer, richer tapestry. The methodology is not a mysterious one—it is simply a matter of delving through the sources, and through the sources used by those sources, finding works bound in with other works, following clues rather than relying on tables of contents and indices.

Gilles Menage wrote a history of women philosophers (Menage 1690) employing a colloquial use of the term "philosopher" to mean "intellectual." In it, he included women gynecologists, astrologers, queens, pharmacists, etc. as well as those whom we would describe as "philosophers." His descriptions were a brief sentence or two. It was necessary to trace down the source materials that he used (and often, *their* sources) until fuller pictures emerged. Some have referred to this as "philosophical archeology," and in some cases, it has been necessary to attempt to track down first editions, recensions of ancient *scholia*, etc., but generally speaking, far less effort is required to "unearth" works by our early foremothers.

If we must choose between the Lazy Boy and the Female Detective methodologies, I'd recommend the latter—unless, of course, you have no interest in the truth, in the TC.

1.3 Whether and How Excluded Women Philosophers Have Been Included in Philosophy

Women philosophers have been included in the histories, epitomies, and earliest encyclopedias of philosophy since antiquity. Lucian includes mention of Diotima of Mantinea, made famous by Plato as Socrates' teacher in the *Symposium* (Lucian, *Portraits.*) Thucydides (Thucydides, *Peloponnesian Wars* 1982) and Plato (Plato, *Menexenus*) preserved nearly identical versions of Aspasia's argument as to why the state does not owe pensions to widows and orphans of soldiers killed in action. Aristotle

(*Poetics* 1911, 1458a24, *Rhetoric*, 1405b) mentions Cleobulina of Rhodes and reproduces two of her logic riddles. Plutarch in *Parallel Lives* mentions ancient women philosophers, Diogenes Laertius in *Lives and Opinions of the Eminent Philosophers* (*Laertius* 1911) appends a list that includes many women philosophers in the Pythagorean tradition: some of their writings and records of their teachings have survived the millennia. The Suda *Florilegium* and *Lexicon* is our only source of Aesara of Lucania's *Book on Human Nature*. These just-mentioned works as well as a number of others came from a period well before the printing press, when works were copied on parchment and papyrus, by hand, at civil, royal and monastic libraries that made a few recensions. I could go on and on—the point is that prior to the seventeenth-century women generally were included in, not excluded from, the most comprehensive of reference materials of earlier epochs.

From the seventeenth century onwards, there were only a few encyclopedists, historians, and epitomists who did include women philosophers, reproducing their texts, summarizing or mentioning them. As mentioned above, Menage produced an early modern history of women philosophers, as did Alexander (1782). At the dawn of the twentieth century, Hermann Diehls reproduced works by Neo-Pythagorean women philosophers (Diehls 1903). These are just a few examples of highly-regarded scholars of this period who preserved contributions made by women to the history of philosophy.

Works by women philosophers also were preserved in source materials that were further removed from philosophy and its history. Catharine Trotter Cockburn's *Defence of Locke* was buried in the published collection of her plays (Cockburn 1702). Oliva Sabuco's work, *New Philosophy of Human Nature*, although republished at least once a century since it burst on the scene in 1587, was preserved amongst the literary works by the government of Spain. Looking outside the discipline was apparently too much to expect of most nineteenth and twentieth century historians of philosophy. But a cursory reading of works by women that have been categorized as belonging to another discipline yields important results. We find philosophical writings throughout the discipline of religion in works by Teresa de Avila (1588) and Zeb-un-Nissa, in physics by Émilie Du Châtelet and Mary Somerville, in mathematics by Hypatia of Alexandria and Mary Everest Boole (1884, 1903). We find philosophy published by Christine Ladd-Franklin in psychology journals and as literature by May Sinclair, as medicine by Hildegard von Bingen and Oliva Sabuco, and published as social work by Catharine Beecher (1836, 1846) and Jane Addams (1907, 1915, 1917, 1922).

Were twentieth-century historians of philosophy like Copleston (1942) and the Durants looking beyond their noses to the great resources of earlier centuries, they would have found the treasure trove of works by women philosophers. Like the *Dead Sea Scrolls*, it is probable that some of these more than a thousand works will turn out to be important works in the TC. Unfortunately, too many professional philosophers who wrote histories of philosophy, who taught the subject, or who served in academic offices where curricular content was decided, took shortcuts rather than report the truth. Is this incompetence? Yes. Is this bigotry? Yes.

1.4 Whether and How Feminist Philosophy and the History of Women Philosophers are Related

Tracy Bowell identifies three distinct but interrelated problems related to the marginalization of feminist philosophy that are relevant to the present discussion: Woman Problem 1 (the scarcity of women in the profession and the masculine image of the professional philosopher that pervades both the image of "a philosopher," and the content of "what is philosophy?"); Woman Problem 2 (the scarcity of works by women in the accepted canon of philosophy and in the content of programs of advanced philosophy education, accompanied by "women's absence from the discipline qua its discourses, the texts, debates, problems, and conversations of which it is comprised"); and (3) the Anti Feminist Problem (the marginalization of feminist work that is legitimately philosophical, and the profession's refusal to grant proper status to gender-conscious and feminist work, particularly in light of its paradoxical acceptance of such work as "marginal" or a "specialty interest" (Bowell 2015, p. 5).

There are several sorts of relationships between feminist philosophy and the history of women philosophers. Those relationships are for the most part, not very interesting, and in part turns on how one defines feminist philosophy. The first sort of relationship is that between authors of feminist philosophy and women who are generally excluded from the HC. If we take a fairly weak definition of feminist philosophy as theories/arguments for women's equality to men, then, some early women philosophers—from antiquity through the twentieth century—were feminist philosophers. They were written out of the modern histories, in part for reasons I have described above. The surviving fragmentary works of these authors, including the Pythagorean Theano, (Waithe *HWP 1*, pp. 12–15) and Neo-Pythagoreans Theano II and Phyntis (Waithe *HWP 1*, p. 141ff.) offer sex-complementarian views of women that hold that some virtues are proper to men, some to women, and some to both. Theano's "*Better to be on a runaway horse than to be a woman who does not philosophize*" is a clear feminist statement about women doing philosophy that suggests that being a philosopher is not reserved for men. Sex-complementarian views also are promulgated in the Medieval period by Hildegard von Bingen in *Liber Vitae Meritorum* (II, 5, IV, 32). We see it in the early modern period through the nineteenth century in arguments proffered by Olympe de Gouges that women's special contribution of citizens to the society entitle them to maternity care even if men could have no similar claim (Gouges, *Droits des Femmes*). The sex-complementarian view is that men's and women's virtues (intellectual, moral, physical) are sometimes different, but always of equal importance to those of men. I will not argue here for or against sex-complementarianism, but clearly, it qualifies as a feminist position: it is a theory that men and women are of equal worth, albeit sometimes in gender-differentiated ways (Allen 1985, *inter alia*).

If, on the other hand, we take a fairly strong definition of feminist philosophy as theories/arguments for women's equality to men, then early modern women philosophers who made explicit political arguments for women to have the same social, legal and civil rights as men (not merely rights that are equivalent to those of men) qualify as feminist philosophers. In this vein, the history of women's contributions to philosophy starts a bit later in the history of our discipline, during the Medieval period with (and this certainly is a matter of opinion) Christine de Pizan as its earliest explicit proponent.

The connection between the history of women philosophers and feminist philosophy goes deeper than the fact that historically, there have always been feminist women philosophers. Feminist concerns about philosophy illuminate the question "why are women philosophers so poorly represented in the histories and reference works in philosophy and in the academic programs that create new philosophy professors?" The answer is along the lines identified by Bowell. A profession that has for recent centuries been dominated by men continues to be dominated by men for several reasons. First, is the assumption that women are less capable of reasoning (and hence, of doing philosophy) than are men. On the assumption that this is true, terms such as "woman philosopher" translate as "dilettante," or "derivative" and writings by women are dismissed by historians out of hand, unread, without being given the consideration that would have been offered to a male author. I've been just as guilty as the historians I've disparaged of dismissing works by women as not philosophical: the title of Margaret More Roper's *Devout Treatise upon the Pater Noster* seemed likely to be a devotional discussion of the Lord's Prayer. Upon examination, it is an astute feminist analysis of resolving conflicting duties of obedience to the King with one's conscientiously-held religious beliefs. This time-honored topic of moral and political philosophy clearly merits Roper's inclusion in TC.

A second reason for women philosophers' under-representation in HC is the dismissal by male philosophers of feminism and feminist philosophy as being philosophy. The entire area is disregarded on the assumption that it is not philosophical. At Bowell notes, the area is described with less derogation, but the same effect, as "specialized," or "marginal." It is dismissed (by men) out of hand because it is likely to be of primary interest to women. This is akin to a Caucasian rejecting a book that analyzes racism as a moral and political problem because it will primarily be of interest to blacks. Both the rejection on the basis of gender and the rejection on the basis of gendered-interest are instances of unscholarly and unphilosophical gender bias.

1.5 Are There Any Themes or Arguments that Are Common to Many Women Philosophers?

Certainly more pre-twentieth century women than men have discussed feminism and indeed have developed feminist philosophy. Fewer pre-twentieth century women than men have written about logic. Many women philosophers also excelled in arts, sciences, and fields of inquiry outside our discipline. These are the only generalizations I can come to on the basis of my own study of the writings of approximately one hundred eighty women philosophers writing prior to 1980. The overwhelming majority of women philosophers have written in exactly the same divisions of philosophy—ethics, metaphysics, epistemology, etc.—as have men philosophers. And they have written on exactly the same topics and issues as their male counterparts. To my knowledge, with one exception (Oliva Sabuco, sixteenth century), all of the women philosophers were participants in circles that included male and female philosophers, and all addressed issues common to their contemporaries. Sometimes, this participation was through personal contact. Sometimes women philosophers engaged in correspondence with male philosophers who were their contemporaries.

1.6 Does the Inclusion of Women in TC Require a Reconfiguration of Philosophical Inquiry?

I hope that the discussion above aptly demonstrates the answer to this question. No, we do not need to reconfigure how we do philosophy, nor what constitutes philosophy. What we do need to do is to recognize that in recent history, philosophers have failed at our most basic task: asking the right questions. Like Nancy Drew and Agatha Christie, we ought not to be satisfied with facile answers. We need to return to the pre-modern practice of preserving texts—and the names and reputations of their authors—based upon the intrinsic significance of those texts, not based upon their marketability or their affordability. As our colleagues in the discipline of Comparative Religion have done with the discovery of the Dead Sea Scrolls, we as a profession need to engage in critical re-examination of these more than a thousand recently-sidelined works. Their importance needs to be considered relative to the state of philosophical inquiry at the time they were written. Their importance needs to be considered on their own merits. Their importance needs to be considered in light of what questions the groups of philosophers with whom their authors communicated—circles they participated in—were discussing. We need to analyze how each individual article or treatise moved the general discussion along, contributed some original insight, raised a significant objection, clarified someone else's idea, was instrumental in refining a colleague's or a student's or a successor's perspectives or led to innovations in related disciplines, etc. When these loose threads are understood and are rewoven in context of the rest of the tapestry that we now believe to be philosophy's TC, then, and only then might we dare think we know the history of philosophy.

What will and now does require reconfiguration is the way in which we teach history of philosophy. We must correct the grievous omissions of past centuries by changing the philosophy curriculum. For the next three or four centuries, we ought to teach only the history of women philosophers.[10] After that, perhaps our successors may wish to integrate the philosophy curriculum along lines of gender. My male colleagues will either laugh heartily at this suggestion or will shriek in alarm without considering that the neglect had been deemed good enough for the goose ought now to be deemed good enough for the gander. However, a commitment to feminist

[10] I owe this suggestion to Professor Janet Kournay of Notre Dame University.

practices in academia need not entail the obliteration of males from our history. What might be politically possible within the discipline is to lengthen the number of courses in the History of Philosophy curriculum sequence from the typical four to six or seven. Just as universities have acknowledged the importance of expanding undergraduate education to include Black Studies and Women's Studies, so, too, ought philosophy programs acknowledge the importance of teaching a history of philosophy that is neither incompetent nor bigoted as is reflected in the early twenty-first-century curricula. Failing to do so is a horribly unphilosophical perpetuation of myth and bias.

References

Addams, J. (1907). *Newer ideals of peace*. New York: Chautauqua Press.
Addams, J. (1915). *The overthrow of the war system*. Boston: Forum Publications.
Addams, J. (1917). *Patriotism and pacifists in war time, City club bulletin* (Vol. 10, p. 9).
Addams, J. (1922). *Peace and bread in time of war*. New York: MacMillan.
Alexander, W. (1782). *The history of women, from the earliest antiquity, to the present time*. London: np.
Allen, P. (1985). *Concept of woman*. Montreal: Eden Press.
Aristotle. (1911). *Collected works, Vol. 11, de Rhetorica and de Poetica*. Oxford: Clarendon Press.
Avila, Teresa de. (1588). *El Castillo Interior*, Salamanca, 1588. (E. A. Peers, Trans.). (2008. *Teresa of Avila, Interior Castle*).
Beard, M. (1942). *Woman as a force in history*. New York: MacMillan.
Beecher, C. (1836). *Letters on the difficulties of religion*. Hartford: Belknap & Hammersley.
Beecher, C. (1846). *Common sense applied to religion, or the bible and the people*. New York: Harper & Bros.
Boccaccio. (1374). *De Claris Mulieribus*. np.
Boole, M. E. (1884). *Symbolical methods of study*. London: K. Paul, Trench & Co.
Boole, M. E. (1903). *Lectures on the logic of arithmetic*. Oxford: The Clarendon Press.
Bowell, T. (2015). The problem(s) of women in philosophy: Reflections on the practice of feminism in philosophy from contemporary Aotearoa/New Zealand. *Women's Studies Journal, 29*(2), 4–21.
Cockburn, C. T. (1702). *The defence of Mr. Locke's essay of human understanding*. London.
Copleston, F. (1942). *A history of philosophy* (Vol. 1). New York: Penguin Press.
De Pizan, Christine. *Livre de la Cité des Dames*. Bibliotheque Nationale de France, *Département des Manuscrits Français, Ancien fonds* 7395.
Diehls, H. (1903). *Die Fragmente der Vorsokratiker*. Berlin.
Du Châtelet, É. (1740). *Institutions de Physique*. Paris.
Du Châtelet, É. (1744). *Dissertation Sur la Nature et la Propagation du Feu*. Paris.
Durant, W & Ariel. (1935–75). *Story of civilization*. New York: Simon & Schuster.
Gouges, O. (1791). *Les Doits de la Femme*. Paris: Bibliothèque Nationale de France E 5588.
Hildegard von Bingen. (2003). *Beate Hildegardis Cause et cure*. In L. Moulinier (Ed.). Berlin: Akademie Verlag.
Hypatia of Alexandria. (1943). Rome, —A. *Commentaires de Pappus et de Theon d'Alexandrie sur l'Almageste, Tome III., Theon d'Alexandrie Comentaire sur les livres 3 et 4 de l'Almageste, Studi e Testi, 106*, Citta del Vaticano: Biblioteca Apostolica Vaticana.
Julian of Norwich. (1978). *A book of showings to the Anchoress Julian of Norwich*. E. Colledge & J. Walsh (Eds.). Toronto: Pontifical Institute of Medieval Studies.
Ladd Franklin, C. (1899). On some characteristics of symbolic logic, *American Journal of Psychology*, *2*, 543–567.

Ladd Franklin, C. (1929). *Colour and colour theories*. London: K. Paul, Trench, Trubner.
Laertius, D. (1911). *Lives and opinions of the eminent philosophers* (R. D. Hicks, Trans.). Loeb Classical Library.
Lucian. (1905). *Portraits [A Portrait Study]* in HW and FG Fowler, *The works of Lucian of Samosata, Volume III*. Oxford: Clarendon Press.
Menage, A. (1690). *Historia Mulierum Philosopharum*. Lugduni. Trans. Beatrice Zedler 1984 as Gilles Menage. In *History of Women Philosophers*, Lanham MD.
Murphey, M. G., & Flower, E. (1977). *History of philosophy in America*. New York: Putnam.
Plato. (2005). Menexenus. In Hamilton (Ed.), *Collected dialogues of plato*. Cambridge MA: Bollingen Press.
Plutarch. (1914). *Parallel lives*. Dryden, transl. Hearst International.
Roper, M. (1526). *Devout Treatise upon the Pater Noster*. London: T. Berthelet.
Sabuco, O. (1587). *Nveva Filosofia de la Naturaleza del Hombre* Madrid: Pedro Madrigal, in English translation as *Sabuco, Oliva, New Philosophy of Human Nature*, Waithe ME, Vintro MEC and Zorita CA, translators. University of Illinois Press, 2007.
Sinclair, M. (2015). *The complete works of May Sinclair*. Kindle edition.
Somerville. (1834–1846). *On the connexion of the physical sciences*. New York: Harper.
Suidae. (1853). *Lexicon*, Gaisford translation Greek and Latin. Berlin: Sumptibus Schwetschkiorum.
Thucydides. (1982). *Peloponnesian wars*. New York: McGraw-Hill.
Wade, I. (1941). *Voltaire and Madame du Châtelet: An Essay on the Intellectual Activity at Cirey*. Princeton: Princeton University Press.
Waithe, M. E. (Ed.). (1987). *A History of Women Philosophers, Volume 1: Ancient Women Philosophers*. Dordrecht and Boston: Martinus Nijhoff/Kluwer.
Zeb-un-Nisa. (1913). *Diwan*. Lal and Westbrook, translators. London: np.
Zinsser, J. (2009). *Émilie du Châtelet: Selected philosophical and scientific writings*. Chicago: University of Chicago Press.

Chapter 2
The Recognition Project: Feminist History of Philosophy

Charlotte Witt

Abstract The question "Why does the history of philosophy matter to feminist philosophy?" is a particular version of the question "Why does the history of philosophy matter to philosophy?" And the latter question opens up into debates over the methods and purposes of those philosophers who engage with the history of philosophy. But, of course, there is more than one way for the history of philosophy to matter. In this paper, I suggest that feminist historians of philosophy who are recovering the forgotten work of women philosophers are engaged in a project, which I call *the recognition project*. Furthermore, I argue that if we focus on the theme of recognition in both its epistemic and political aspects, we will be motivated to add "feminists" to the categorization of scholarly work in the history of philosophy.

2.1 Introduction: Why Does the History of Philosophy Matter to Feminist Philosophy?

The question "Why does the history of philosophy matter to feminist philosophy" is a particular version of the question "Why does the history of philosophy matter to philosophy?" And the latter question opens up into debates over the methods and purposes of those philosophers who engage with the history of philosophy. It turns out that there is more than one way for the history of philosophy to matter. A recent book on early modern philosophy, *Philosophy and Its History: Aims and Methods in the Study of Early Modern Philosophy*, provides the following typology of the aims and methods of work on the history of philosophy:

Thanks to Ásta Sveinsdóttir for suggesting this title for my interpretation of feminist scholarship on the history of philosophy, and to the audience at *Exploring Collaborative Contestations and Diversifying Philosophy Conference*, Villanova (5/15) for useful discussion.

C. Witt (✉)
Department of Philosophy, University of New Hampshire, Hamilton Smith Hall Rm 249P, Durham, NH 03824, USA
e-mail: charlotte.witt@unh.edu

S. Thorgeirsdottir and R. E. Hagengruber (eds.), *Methodological Reflections on Women's Contribution and Influence in the History of Philosophy*, Women in the History of Philosophy and Sciences 3, https://doi.org/10.1007/978-3-030-44421-1_2

(a) the history of philosophy is held to be a source of ideas and arguments that may be of use in contemporary philosophy ("appropriationists")
(b) the history of philosophy is to be studied and understood for its own sake and in its own terms, even when the problems of interest to the figures in this history have since fallen off the philosophical agenda ("contextualists")
(c) Continental philosophy, in which one's philosophical position is developed dialectically with a tradition that is often constructed simultaneously for that purpose. (c) is version of (a) except that (a) is not committed to the idea that philosophy is an intrinsically historical discipline ("continentalists").[1]

The first two types of history of philosophy ("appropriationist" and "contextualist") are familiar from other discussions, although the terminology used to label them varies. For example, Cynthia Freeland contrasts two models of the history of philosophy: philosophical history of philosophy (= appropriationist) and exegetical history of philosophy (= contextualist).[2] The editors recognize a third type of history of philosophy ("continentalist") in which one thinks against and within a tradition that is—to a greater or a lesser degree—a product of that dialectical engagement. With this addition, the taxonomy of methods and aims appears more complete and adequate than those that include only the first two types, which have dominated discussions of the nature and value of the history of philosophy in the Anglo-American tradition.[3]

But, where should we place feminist work in the history of philosophy? Does it fall neatly into one (or perhaps more than one) of the three alternatives? In this paper, I suggest that feminist historians of philosophy who are recovering the forgotten work of women philosophers are engaged in a project, which I call *the recognition project*, and I argue that this project should motivate a fourth category, "feminists". I argue further that recognition, which has both an epistemic and a normative component, is common to feminist work that might otherwise be classified as appropriationist, contextualist or continentalist. If we focus on the recognition project, in both its epistemic and political aspects, we will be motivated to add "feminists" to the categorization of scholarly work in the history of philosophy.

As a preliminary, it is useful to review the large body of work that has been done over the past 25 years and falls under the umbrella of the feminist history of philosophy. I consider the extent to which the threefold division fully captures feminist work in the history of philosophy pointing out where it seems to me not to be a good fit. I focus, in particular, on recent scholarship on women philosophers in the early modern period to show the inadequacy of the threefold categorization, which does not recognize the unique character of this philosophical and scholarly work. I develop the notion of recognition to capture and to express both the epistemic and the political aspect of the feminist work on the history of philosophy and to flag

---

[1]Laerke et al. (2013), 1, 3.

[2]See also Freeland's important discussion of feminist work in the history of philosophy in "Feminism and Ideology in Ancient Philosophy" (2000).

[3]For an interesting and influential analysis that questions the opposition between appropriationists and contextualists in the Anglo-American tradition, see Wilson (1992).

the parallel absence or inadequate representation of feminist work in mainstream methodological reflections on the history of philosophy. Taxonomies of historical scholarship that do not recognize the unique task of feminist historical scholarship reiterate the lack of inclusion of women philosophers in the history of philosophy. The act of recognition of the unique character of feminist work in the history of philosophy is both an intellectual act of seeing what had not been previously visible and a normative act of recognizing that the invisibility is unjust. It is, therefore a categorization that runs orthogonal to mainstream typologies of work in the history of philosophy in terms of methods and goals. As is often the case with feminist contributions to philosophy feminist work in the history of philosophy both destabilizes mainstream categories and often goes unrecognized.

2.2 Feminist History of Philosophy

I divide feminist work in the history of philosophy into four types[4]:

1. Criticisms of the canon (or individual canonical philosophers) as misogynist
2. Appropriations of canonical philosophers
3. Revisions to the history of philosophy
4. Methodological reflections.

How does this way of categorizing feminist work in the history of philosophy relate to the methodological typology of appropriationalists, contextualists, and continentalists? There is apparently considerable overlap. For example, we might pair up "Appropriations of canonical philosophers" with appropriationists, thinking here of feminist philosophers who advocate for the usefulness of canonical philosophers and their theories for various feminist purposes.[5] And the revisions to the history of philosophy that consists of retrieving, editing, translating, and publishing the work of "unknown" women philosophers, it might seem, is a contextualist project. And finally some of the feminist work that falls under the first category, the critique of a canonical figure or period of philosophy as a misogynist, it seems to me, most closely resembles the approach of continentalists, who engage critically with the canon (or with individual canonical philosophers) to create and develop it anew.

It seems, then that feminist work in the history of philosophy can be categorized—perhaps without remainder—using the triad of appropriationists, contextualists, continentalists. It seems that there is nothing characteristic of the feminist history of philosophy that would warrant or ground a separate category. In this paper, I develop the notion of recognition as providing the thread that unites the diverse projects that go under the label of feminist history of philosophy.

[4]See "Feminist History of Philosophy", *The Stanford Encyclopedia of Philosophy*.

[5]For a sampling of this kind of feminist scholarship (among other kinds and perspectives) see the Penn State Press series *Re-Reading the Canon* and my review of the series "Feminist Interpretations of the Philosophical Canon" (2006).

2.3 The Recognition Project

In social and political philosophy recognition is a term used to refer to a normative relationship between two individual human beings, although it is also used for relationships among groups. In this usage recognition has epistemic connotations; to recognize someone is to see that person in a particular way. I could recognize a female graduate student as a fellow philosopher for example. Now, I could simply be wrong about what that graduate student is studying which would be a misrecognition that is only an epistemic error. However, misrecognition of a person can be both an epistemic mistake and political harm. To misgender a trans person, for example, is not merely a classification error but it is also a harm to that person. What I want to suggest is that the notion of recognition, retaining its double epistemic and political meaning, can be usefully applied to feminist work in the history of philosophy; it is a thread that unites otherwise apparently disparate projects. Recognition, as I will use the term, is the acknowledgement of something's existence or its validity (or both). Hence, it is a term with intertwined epistemic and normative aspects. Recognition describes both the epistemic act of noticing that something exists and the granting of validity or worth to that which has been acknowledged to exist. Both aspects of recognition are relevant to feminist work in the history of philosophy as it focuses on questions of existence and questions of validity or worth. Since feminist work on the history of philosophy of various kinds shares the goal of recognition it stands alone, and cannot be simply classified as appropriationist, contextualist, or continentalist, without remainder. At another level, the lack of recognition of the feminist history of philosophy, or its misrecognition, reiterates the original epistemic error and normative harm that the recognition project addresses.

The clearest example of the recognition project is the work that identifies women philosophers and retrieves their work for the historical record. It is a project that argues not only for the existence of women philosophers but also for their worth as philosophers. Hence it is an epistemic project with a normative dimension and for this reason, it cannot simply be classified as contextualist—even though much of this work is concerned to place the women philosophers in their historical context and to understand their ideas in relation to their philosophical and cultural milieu. Nonetheless, I will argue that it is not *simply* contextualist history of philosophy and that to classify it as such is an instance of misrecognition.

Before continuing I want to underline that my conception of feminist history of philosophy as a recognition project does not presume to capture the intentions of the participants, which might vary. In particular, in claiming that feminist work in the history of philosophy can be usefully interpreted as a project of recognition, I am not committed to the view that participants would see their scholarship in this light. Rather, I am suggesting that certain features of what would otherwise be quite diverse projects can be usefully categorized as part of a single recognition project and that we would leave out something important about this work if we simply shoe-horned it into the three mainstream categories.

2.4 Revisions to the History of Philosophy: Beyond Contextualism

I begin with the most obvious case. Feminist history of philosophy is most distinctive, and most radical, in its retrieval of women philosophers for the historical record, and in its placement of women in the canon of great philosophers. The latter task raises difficult questions, which I am not going to address: What is a philosophical canon? How is it formed? What is its purpose? Instead, I focus on the exciting and substantive research on women philosophers in the early modern period. There is a large (and growing) body of careful historical scholarship documenting the presence and importance of women philosophers in the early modern period and making their views available.[6] It is striking that this research is not mentioned in the discussion of methods and goals in the history of philosophy that yields the division into contextualists, appropriationists, and continentalists.

Perhaps the recent scholarship to identify women philosophers and their writing in the early modern period falls neatly into the contextualist camp. There is reason to think so. According to Lisa Shapiro: "[Rather] scholars aim (1) to make long out-of-print texts accessible again; (2) to develop interpretations of these texts that (a) bring out their philosophical content and (b) demonstrate the involvement of these women in the philosophical debates of the period."[7] Is this a contextualist project? It might seem so since it requires a careful study of texts in their cultural context and a focus on philosophical debates of the period that may have "fallen off the contemporary agenda". However, there is also reason to think that this description, while accurate to the scholarly methodology, does not fully capture the scale and significance, and hence the aim, of the project. To see this we need to consider it in its historical context.

First, consider that *The Encyclopedia of Philosophy*, published in 1967, which contains articles on over 900 philosophers, includes articles on only three women: George Eliot, Madame de Staël and Saint Teresa of Avila; two novelists and a saint. And, lest you think that the list of 900 includes only philosophical heavy hitters, the editor tells us: "We have also made it a point to rescue from obscurity unjustly neglected figures, and in such cases, where the reader would find it almost impossible to obtain reliable information in standard histories or in general encyclopedias we have been particularly generous in our space allotments" (x). In that effort, not a single woman philosopher was considered worthy of an entry. The world of the 1967 *Encyclopedia of Philosophy* is one in which there literally were not any women philosophers of any note. One might infer that there simply were no women philosophers who were "unjustly neglected figures". But, of course, there were. Indeed, earlier catalogues and lists of philosophers include the names of women.[8]

---

[6]Project Vox is a good example. http://projectvox.library.duke.edu.

[7]"Feminist History of Philosophy", *Stanford Encyclopedia of Philosophy*.

[8]For a discussion see Shapiro (2004).

The 2005 *Encyclopedia of Philosophy* records the beginning of a sea change with 35 entries for women philosophers including important figures from the early modern period. What I want to argue is that the "discovery" of women philosophers in the history of philosophy is both an epistemic and a normative project. The epistemic part is congruent with a contextualist methodology, but the normative aspect is both antithetical to the self-understanding of contextualist interpretation of the history of philosophy and not reducible to it.

First, let's consider the irreducibility point. Recall that recognition is both an epistemic and a political act in that it acknowledges both the existence and the validity of a subject. It is in making a case for the validity of women philosophers of the early modern period that we can see the political aspect of this scholarship most clearly. The political aspect of recognition, in turn, has at least two components. First, there is the problem of misrecognition. As an example of misrecognition consider that there is often push back to the status of these women *as philosophers*. For example, consider the debate between Alanen and Garber over whether Elizabeth of Bohemia was a philosopher or a "learned maid".[9] Those working on women philosophers in the early modern period face the threat of misrecognition of their subjects and have to address it either directly or indirectly. The second political component is that these philosophers were overlooked in encyclopedias and canon formation not randomly but because they were women.[10] And that is unjust. Hence, a project devoted to their recovery is inherently political; it addresses an epistemic injustice. The same tasks would not face a historian of philosophy who is rehabilitating a minor male figure of the early modern period. The political aspect of the recognition project is not reducible to, or detachable from, its epistemic component.[11]

There is also an inherent tension between the contextualist methodology and the normative aspect of the recognition project. The contextualist, who only considers the philosopher in his or her historical context, would not consider it compatible with her methodology to address issues of misrecognition nor would her work make assumptions about exclusion and epistemic injustice. These normative and political assumptions import concepts that might not be native to the philosopher's historical and intellectual ecology.[12] Viewed as a recognition project the proliferation of scholarship on women philosophers in the early modern period is not adequately classified as a contextualist project without remainder.

But now it is reasonable to wonder how the notion of recognition applies to feminist historians who want to appropriate useful ideas from canonical figures like Aristotle or Kant or, alternatively, who are critical of the misogyny in a philosopher's concepts or, indeed, in the Western philosophical tradition as a whole? These projects

---

[9]Lilli Alanen "Descartes and Elizabeth: A Philosophical Dialogue?" in *Feminist Reflections on the History of Philosophy* (Alanen 2004; Alanen and Witt 2004).

[10]For a useful discussion of these issues (though not in the context of the concept of recognition) see Lisa Shapiro's "Some Thoughts on Women's Place in the History of Philosophy" (2004).

[11]Although this paper considers feminist work on the history of philosophy as a recognition project, I believe that other attempts to diversify philosophy can also be understood as recognition projects.

[12]It is important to note, however, that some early modern women philosophers joined in a contemporary debate over the status of women in society. See Shapiro (2004).

seem at odds both with one another and with feminist contextualism. How do they differ from mainstream appropriationists and continentalists? Do they have anything in common? And, in what way do they fit into the recognition project? What is recognized to exist? And what is deemed legitimate?

2.5 Appropriation, Critique, and Recognition

I turn first to feminist work in the history of philosophy that urges appropriation of the ideas and theories of canonical philosophers that might be put to feminist purposes. How does this scholarship differ from mainstream appropriationists, who use the history of philosophy as a source of ideas and arguments that may be of use in contemporary philosophy? Does feminist appropriation of historical figures and ideas simply fit into this category without remainder? And, if not, in what sense can they be characterized as part of a recognition project? Recall that recognition has both an epistemic and a normative/political component. I will argue that feminist appropriations of traditional philosophy have these two aspects as well and in this way cannot fit into the appropriationist category without remainder.

Initially, it makes sense to categorize those feminists who work on the history of philosophy in order to retrieve resources useful for contemporary feminist theorizing as appropriationists. But they differ from mainstream appropriationists in at least one important respect. The issue of appropriation of traditional philosophical ideas for use in feminist theory is contested—not simply in particular cases—but in general. For example, if the ideas are bound up in a philosophical perspective that is misogynist there is a problem of extracting the valuable ideas and the problem of whether or not they are genuinely useful for feminist purposes. As Cynthia Freeland notes "The inheritance perspective can be seen to perpetuate oppressive relations insofar as it seeks legitimacy for feminists now by taking possession of a tradition that we should perhaps not be so quick to claim".[13] In addition to the epistemic task of showing how and in what ways a particular set of ideas from the history of philosophy might be useful for feminist theory, the feminist appropriationist faces an additional political challenge to the entire enterprise. The defense of the utility of the history of philosophy for purposes of contemporary feminist theorizing is double; it is both epistemic and political. Hence feminist appropriation of the history of philosophy does not fit into the appropriationist category without remainder.

It is worth noting that the task facing feminist appropriation of the tradition differs in type and in scope from an apparently similar challenge facing historians of philosophy with regard to an individual philosopher's unacceptable political views, for example, Frege's or Heidegger's anti-Semitism. The difference in type is that would-be appropriationists of Frege or Heidegger have available to them a norm of interpretation that permits segregation of a thinker's mere political opinions or personality from his philosophical views. The difference in scope is that feminist

[13] Freeland (2000), 389.

appropriators face a long and complex history of exclusion, which makes the strategy of segregation difficult and contested.

The continentalist appropriation of the history of philosophy differs from mainstream appropriation by being both a dialectical engagement with the tradition and imbued with the belief that philosophy is, in an important sense, an historical discipline.[14] How well does feminist work in the continentalist tradition fit this description?

I will use as my example a recent paper on how to address the sexism and racism in Kant's philosophy by Dilek Huseyinzadegan:

> Therefore, I argue that the best thing that we can hope to achieve as feminist appropriators of canonical figures such as Kant is to employ a methodology of constructive complicity [inspired by Gayatri Spivak's *A Critique of Postcolonial Reason*]. First, we must admit that we as professional philosophers constructing and re-constructing Kantian arguments are *complicit* in the problems that Kant's texts exemplify. Then, we must highlight and inherit these problems as our own issues rather than disavowing them as the historical limitations of the man himself. … From such a position of complicity, we can then hope … to offer more nuanced and honest interpretations of his work as well as our philosophico-political challenges today.[15]

Here, we find expressed both the dialectical character of continentalist appropriations of canonical figures and the view that philosophy is intrinsically a historical discipline. Our philosophical world is, at least in part, an inheritance from Kant—warts and all. But what should we include in that inheritance? Should we "white out" Kant's racism and sexism (for example, by segregating his pre-critical from his critical thought). Should we simply ignore these features of his thought and resurrect a "good" Kant? How do we make Kant's texts our inheritance?

In my earlier discussion of feminist appropriation of canonical philosophical texts and how it differs from mainstream appropriationism I suggested that feminist inheritance and use of traditional philosophical concepts is always problematic in a way that mainstream appropriation is not. How can we be sure that our use of a philosopher's ideas is not implicated in his objectionable sexism (or racism) expressed elsewhere? How can we be sure that what looks like a purely formal concept does not have hidden political content and implications? Here we find a different suggestion, one that underlines the historical character of philosophical thinking for continentalists. For a continentalist feminist, it is precisely the philosopher's sexism or racism that we need to understand as our inheritance, not what we need to expurgate from the text. "I suggest that we redirect our focus from what is salvageable (the good Kant or Kantianism) to inheriting Kantianism as a whole and that we make Kant's texts ours (i.e., feminists] by reclaiming them in their entirety, the good and the bad (and

---

[14]For one explanation of the intrinsic historicity of philosophy (including contemporary philosophy) from the continental perspective, see Robert Scharff's *How History Matters to Philosophy: Reconsidering Philosophy's Past After Positivism* (Scharff 2014; Witt and Shapiro 2018).

[15]Huseyinzadegan (2018), "For What Can the Kantian Feminist Hope? Constructive Complicity in Appropriations of the Canon."

the ugly)."[16] Here again, it is the explicit political orientation of the feminist continentalist inheritance that differentiates her interpretive stance towards the history of philosophy from that of her continentalist brethren.

2.6 Conclusion: No More Disappearing Ink

Eileen O'Neill's groundbreaking 1998 essay "Disappearing Ink: Early Modern Women Philosophers and Their Fate in History" provides a rich and detailed account of women philosophers of the period, and a complex explanation of their disappearance from the history of philosophy. Using the notion of recognition, a concept with both an epistemic and a normative aspect, I have argued that mainstream taxonomies of methods and aims in the history of philosophy should add the category "feminists". Why is this addition important? Why not simply shoehorn the various types of feminist work on the history of philosophy into the existing mainstream taxonomies and ignore any evidence of bad fit? For the very same reason that acknowledging the existence and validity of women philosophers in the history of philosophy is important. It exists, its aims and methods are valid, and it deserves recognition (Kourany 1998).

References

Alanen, L. (2004). Descartes and Elizabeth: A philosophical dialogue? In L. Alanen & C. Witt (Eds.), *Feminist reflections on the history of philosophy*.

Alanen, L., & Witt, C. (Eds.). (2004). *Feminist reflections on the history of philosophy*. Dordrecht: Kluwer Academic Publishing.

Bourchert, D. (Ed.). (2005). *The encyclopedia of philosophy* (2nd ed.). New York: Macmillan.

Edwards, P. (Ed.). (1967). *The encyclopedia of philosophy*. New York: Macmillan.

Freeland, C. (2000). Feminism and ideology in ancient philosophy. *Apeiron, XXXIII*(4), 365–406.

Huseyinzadegan, D. (2018). For what can the Kantian feminist hope? Constructive complicity in feminist appropriations of the Canon. *Feminist Philosophy Quarterly, 4*(1). Retrieved October 29 2018, from https://ir.lib.uwo.ca/fpq/vol4/iss1/3/.

Kourany, J. (1998). *Philosophy in a feminist voice*. Princeton: Princeton University Press.

Laerke, M., Smith, Justin E. H., & Schliesser, E. (Eds.). (2013). *Philosophy and its history: Aims and methods in the study of early modern philosophy*. Oxford: Oxford University Press.

O'Neill, E. (1998). Disappearing ink: Early modern women philosophers and their fate in history. In J. Kourany (Ed.), *Philosophy in a feminist voice*.

Shapiro, L. (2004). Some thoughts on the place of women in early modern philosophy. In L. Alanen & C. Witt (Eds.), *Feminist reflections on the history of philosophy* (pp. 219–250).

Scharff, R. (2014). *How history matters to philosophy: Reconsidering philosophy's past after positivism*. New York: Routledge.

Wilson, M. (1992). History of philosophy in philosophy today; and the case of the sensible qualities. *Philosophical Review, 101*(1), 191–243.

[16] Huseyinzadegan (2018).

Witt, C. (2006). Feminist interpretations of the philosophical canon. *Signs: Journal of Women in Culture and Society*, *31*(2).
Witt, C., & Shapiro, L. (2018). Feminist history of philosophy. In E. N. Zalta (Ed.), *The Stanford encyclopedia of philosophy* (Fall 2018 Edition). https://plato.stanford.edu/archives/fall2018/entries/feminism-femhist/.

Chapter 3
"Context" and "Fortuna" in the History of Women Philosophers: A Diachronic Perspective

Sarah Hutton

> The history of philosophy will be written many more times in terms of emerging themes and theories; in terms of the interests and concerns of thinkers yet to come; and in terms of new information and insights about the past. (Popkin 1999, p. xvii).

Abstract This paper explores the idea of philosophical *fortuna* (fortune) in relation to the study of women philosophers of the past. Tracing a philosopher's *fortuna* involves examining the reception of her philosophy across time in the context of changing philosophical debates, and the reworking of philosophical ideas in new and different circumstances. I argue that focusing on philosophical *fortunae* offers a means to reassess the realities of exclusion and marginalization of women philosophers, and to strengthen their claims to be admitted as full members of the community of philosophers. I conclude with a consideration of Mary Hays's *Female Biographies* to this process.

Much of the work on women philosophers in the last 15 years has been a work of recovery. That work has borne fruit both in terms of the number of female philosophers that have been brought into focus and in terms of a better understanding of their philosophy and their own philosophical priorities. The success of this enterprise is largely owing to the historical approach to philosophy adopted by women historians of philosophy. One might even call it a kind of *archaeology*, since it has involved finding out who they were, locating them, exhuming them from oblivion and restoring them to view. Integral to this process, has been a knowledge of the philosophical and historical context in which these women philosophized. We now have a clearer idea of the philosophical standing of women philosophers of the past in their own day, and of who read them subsequently. Nevertheless, we still confront the problem of strangeness and difference in the thought of women from other periods and their relevance to us now. How to make sense of what we find in their work is not always

S. Hutton (✉)
34 Church Vale, London N2 9PA, UK
e-mail: sarah.hutton@york.ac.uk

S. Thorgeirsdottir and R. E. Hagengruber (eds.), *Methodological Reflections on Women's Contribution and Influence in the History of Philosophy*, Women in the History of Philosophy and Sciences 3, https://doi.org/10.1007/978-3-030-44421-1_3

obvious or easy. This applies just as much to women philosophers of the recent past as to those of the remote past. And we still know all too little about what Eileen O'Neill called the "fate in history" of women philosophers. So, this is a good moment to pause for review and to consider methodological issues. In this paper, I shall explore the idea of philosophical *fortuna* as a means for confronting these problems. I shall argue that focus on philosophical *fortunae* has the potential to bring the history of women philosophers full circle with the present; it offers a means to reassess the realities of exclusion and marginalization, and a means to strengthen the claims for women to be admitted as full members of the community of philosophers.

The *fortuna* of a philosopher or philosophy is traceable in the philosophical debates, the re-workings of ideas and arguments, the reapplication of philosophical ideas in new, and different circumstances that occurred across time. To investigate the *fortuna* (fortune and misfortune) of a philosopher or philosophy involves consideration of the *philosophical* reception of ideas and arguments. It requires examining the take-up (or not) of their philosophical views, and their standing among philosophers at any particular time. And it entails some consideration of the social, cultural, and political conditions associated with that. To trace a *fortuna* is an approach that is found in the history of Renaissance philosophy, especially as practiced in Italy. It is easy to see why it is particularly relevant to Renaissance philosophy, much of which emerged from the recovery and reapplication of classical philosophy, or the re-interpretation of traditional philosophy in the light of new scholarship and changed cultural conditions. Notable instances of this are the transformations that came about through the re-invigorating of Aristotelian tradition, the revival of ancient skepticism, and the Christianization of Platonism. In this paper, I shall consider the applicability of the notion of *fortuna* to the history of women philosophers. Since historical context is a key element in establishing the *fortuna* of a particular thinker or set of ideas, I propose first, to re-assert the claims of *context* for the study of women philosophers, and to review the problems that contextual and non-contextual study of women philosophers raises.

One advantage which we have today over the pioneers in the history of women philosophers is that there is now a pool of research on why we are able to draw.[1] This research has recovered the work of a solid core of philosophical women. No longer do we face skepticism from doubters, who queried whether they counted as philosophers.[2] Whereas previously, much of the work was done by individual scholars working in isolation, we now have the benefit of working collectively—whether we do so consciously or not. Anyone working in this field is now able to draw on other people's research and to derive support and encouragement from their findings and their enthusiasm. In particular cases, there is even enough material available to make it possible to *debate* issues and interpretations. Most of the work which has been done has been on early modern European philosophy, most of it on British women philosophers. But work is beginning to be done on female philosophers from other parts of Europe and on philosophers of the eighteenth century. We can look

[1]Pioneers in the field were Waithe ed. (1987–1995), Dykeman ed. (1999), O'Neill (1998).
[2]Alanen (2004).

forward to the extension of the domain to all periods and across countries. This is hugely exciting, for many reasons, one being that the greater numbers of women philosophizing in the nineteenth and twentieth centuries mean that there is potential to establish a critical mass of women philosophers. Many of the methodological approaches which we employ were established in order to deal with thinkers whose context and concerns are at a distance from our own. Within the wider chronological field, the question must arise as to whether these methodologies apply to the women philosophers who are closer in time and interests. The present paper will have bearing on that, and I shall start by saying that, in my view, the study of modern women philosophers should retain the historical methodologies which have been employed so successfully for dealing with earlier philosophers. In particular, we should not commit the transparency fallacy, by assuming that greater proximity in time means that nineteenth- and twentieth-century women philosophers were philosophically close to us, with similar interests and patterns of thought. The history of women's philosophy will be enormously enriched if we hold on to the idea that the past, even the recent past, is a foreign country, and to research, it requires historical methodologies. The same principle applies to philosophies from different cultures. We cannot study them simply by subjecting them to contemporary Anglo-American philosophical analysis. We need to find a way to understand them in their own terms. And this, in turn, points to the importance of context and the usefulness of the lessons of history.

Now, it is certainly the case that modern interests are driving the recovery of women philosophers. Without the upsurge of interest in women of the past which has been generated in recent years by modern feminism and the women's movement, women philosophers would remain concealed in the dust of history. However, many of the early attempts to discuss women philosophers of the past stalled because of the mismatch between modern philosophical viewing frames and the female philosophers of the past. The failure of progress in these cases is attributable in part to limited historical awareness. This in large part reflects the state of the history of philosophy at that time, which was such, that even what was known about the mainstream philosophy of their times was extremely limited, even misleading. In the early stages of the enterprise of recovering women philosophers of the past, pioneer feminist historians often did not have the information about the circumstances in which their subjects philosophized. A lot of the early work done on women's philosophy of the past was not free from attempts to read modern ideas back into earlier thinkers, to recreate them in our own image, as it were. Feminists were motivated by a wholly understandable desire to recover the predecessors (foremothers) who shared their outlook. Feminist philosophers sought to highlight philosophical themes particular to women in the hope of reconstructing thereby a woman's tradition in philosophy. This was often accompanied by an equally understandable desire to avoid intellectual structures that subordinated women, leading many feminist philosophers to posit female epistemologies. Such approaches often were based on the presupposition that women think differently from men, or, they led to that conclusion. One result was a widespread negative view of rationality, which was often construed as masculine.[3] Now it is perfectly understandable that feminist historians should latch on to those aspects of women's lives and views which speak most directly to their own, or which can be viewed in terms of current theory. But a contemporary perspective which judges the past by the values and assumptions of the present comes at the expense of historicity, produces readings which are flawed because unhistorical. Many a woman philosopher fell outside the "viewing frame" of these early attempts to recover them. Some of them did not seem to be doing philosophy at all. These researches produced considerable theorizing about the paucity of women in philosophy—explaining this in terms of the "masculinity" of philosophy, its overemphasis on disembodied reason, and its evasion of body-centered patterns of thought favored by feminists.[4] And there was a general tendency to focus on the problem of exclusion of women from philosophy.

The historiography of philosophy itself has moved on in the last 25 years, as historians of philosophy began to pay more attention to the context in which their subjects operated. This has required finding out about so-called "minor figures" of philosophy who fall between the high peaks of the *soi disant* "great philosophers". And it has required consideration of the circumstances in which those philosophers considered to be "great" were judged important in their time.[5] This contextual approach to the history of philosophy applies to male philosophers of the past as much as to their female peers. To the extent that such historical and contextual approaches recognize the "otherness" of past philosophers and attempt to address past philosophy in its own terms, they have proved wholly beneficial for the recovery and study of women philosophers of the past. By adopting such a historical approach, a group of feminist historians of philosophy, among whom I would count myself, has made possible the recovery of many figures who were formerly lost to view and established the ground base for discussion of their contributions to philosophy.[6] These historians of philosophy have tackled the problem of where women's philosophy stood in relation to what has been taken to be philosophy as a whole. They did so by reconstructing not just their arguments, but the *context* in which their subjects worked. Such research has enabled us to revise many assumptions and put a stop to many prejudices. Among these were views such as that women did not produce any philosophy of significance, or that they necessarily must have done so differently from men, or that philosophy was misogynistic by definition. The net result of all this work is that it is now possible to discuss women's philosophy without assigning its minor status. The combined results of such historical work are forcing a re-consideration of the

[3]See e.g. Spender (1982), Schiebinger (1991), Bordo (1987), Lloyd (1984).

[4]See *inter alia*: Gilligan (1982), Code (1982), Harding (1991), Gatens (1991), Alcoff and Potter (eds.) (1993).

[5]Notable examples are *The Cambridge History of Renaissance Philosophy*, ed. Schmitt et al. (1988), and *The Cambridge History of Seventeenth-Century Philosophy*, ed. Garber and Ayers (1998). For more adventurous new history of philosophy see Laerke et al. (eds.) (2013), Condren et al. (eds.) (2006). And for feminist history of philosophy, see Alanen and Witt (eds.) (2004).

[6]See for example Broad (2002), Green (2014), Hagengruber (ed.) (2012), Hutton (2004), Shapiro translation of Princess Elisabeth of Bohemia (2007).

narrative of philosophical history, and hence the teaching canon. I would therefore argue (and have argued) that historical research has been essential not just for the recovery of women philosophers, but to enable us to discuss them meaningfully in terms that make sense to us, without diminishing the alterity of past philosophy, neither women's philosophy nor men's.

Nevertheless, emphasis on the historical, and the contextual has its own attendant difficulties. A major problem which such approaches are liable to create is that they serve to underline how different were the thought and concerns of women philosophers of the past from present thinking and concerns, often puzzlingly so. Eileen O'Neill, for example, called attention to both the respective merits of historical reconstruction and rational reconstruction and also the tensions between them. She pointed out that the former highlights the philosophical concerns of the past on their own terms, while the latter focuses on the concerns of the present to be found in the past.

> ...if we utilize the method of historiography known as 'historical reconstruction', we will take as central those issues deemed by the philosophers of the past to be the central ones .... On the other hand, if we utilize the method of historiography known as 'rational reconstruction', the issues philosophically central in a past era will be those that most closely match our current philosophical concerns or that have caused us to have our current concerns.[7]

Thus, a historical approach, too, creates its own "mismatches" between past and present. Although not concerned with women philosophers, Bernard Williams recognized the same difficulty, an often a baffling mismatch between philosophy in the past and the interpretation of it in the present (though he called a historical approach "history of ideas"): "the best possible *history of ideas* is likely to show that philosophy did not in fact mean in contemporary terms what subsequent philosophy has most made of it."[8] Bernard Williams's solution was simply to ignore history, "to neglect or overlook, to some extent, the history that lies between that philosophy and the present day, and to reconsider the philosophy in partial independence from its actual influence". This strategy would impact negatively on the study of women philosophers, who have had more than their share of being ignored! And, if we ignore history, we return to the original problem with which I began, that the recovery and interpretation of women philosophers of the past cannot progress without a historical approach. Reconciling the mismatch between the alterity of past concerns and the interests of the present is another of the challenges which face us. How do we mediate between the productiveness of these respective approaches and the tensions between them? Is it possible to square the two, accommodate the *difference* of past philosophy? I want to argue that it is.

Indeed, I would go as far as to claim that there is a virtue to be made of the awkward problem of these "mismatches". The divergence between the present and the past, be this between reconstructionists of different stripes, or between contemporary theory and past actuality, highlights the fact that the reception of dead philosophers is

---

[7] O'Neill (2007), p. 20.

[8] Williams (2007). Williams first formulated this distinction in the Preface to *Descartes: The Project of Pure Enquiry*. London: Routledge (1978). For further discussion see Hutton (2015).

not historically neutral, but changes and develops over time, in relation to the philosophical cultures within which they are received. The work of recovering female philosophers has now been underway sufficiently long for us to begin to see that even within the short time-span of the last quarter of a century, present interests and theoretical frameworks change. Without undervaluing the work of the pioneers to whom I referred earlier, it is now possible to see more clearly that their readings were *of their time*. A case in point is Carolyn Merchant's *The Death of Nature* (1980), a pioneering work that helped put Margaret Cavendish and Anne Conway in the frame. This work belongs to a decade when feminists posited female epistemologies as they tried to free themselves from intellectual structures that subordinated women. The context of the book was the rise of ecofeminism, which colored feminist interpretations of philosophical reason. In her book, Carolyn Merchant identifies characteristically female ways of thinking about the natural world in the early modern period.[9] This leads her to posit an antithesis between what she saw as older vitalist and holistic views of nature, which she held to be non-exploitative, and the new "mechanical" natural philosophies as represented by Descartes, Hobbes, and the Royal Society. She took the mechanical model to be entrenched, dominant, and domineering, displacing the kind of knowledge and patterns of enquiry to be encountered in Hermeticism and alchemy, which she took to be more female-friendly. As it transpired, not only was Merchant mistaken about the status of the philosophies of the *soi disant* arch-mechanists (Descartes and Hobbes) among women thinkers (e.g., the debt of Anne Conway to Descartes, and Margaret Cavendish to Hobbes), but she overlooked the tensions within science itself, and the tentative and controversial standing of the experimental method of the Royal Society. Merchant mistakenly identified Conway's mentor, Henry More, with mechanists, as an enemy of vitalism. It is entirely questionable that Hermeticism and alchemy were any less exploitative of nature than experimental science, or any less misogynistic. Merchant's analysis is of its time. Its limitations reflect both the limits of historical scholarship at the time, but also the impact of what were then new theoretical frameworks, ecology, and ecofeminism. As an analysis of its time, it exemplifies how the concerns of a later age impact on the reception of the thought of previous eras. Despite my criticisms of a lot of Merchant's work, I recognize that it was borne of a real attempt to engage with the thinking of early women philosophers. It is unquestionably an advance on the judgments of earlier feminists, notably Virginia Woolf, who dismissed Margaret Cavendish's philosophy as "futile", and characterized her as the "harebrained, fantastical Margaret of Newcastle", who "frittered her time away scribbling nonsense". There is a lot else we don't know about Cavendish's reception, or haven't bothered to find out—e.g., the positive assessments of her contemporary, Bathsua Makin, by the essayist Charles Lamb, or the philosopher Mary Hays (on whom more anon). Carolyn Merchant's analysis is part of the modern reception history of Cavendish

[9]Merchant (1980). Merchant deserves credit for a greater degree of historical specificity than was customary at that time, particular, for trying to take account of currents of thought unfamiliar in philosophical history (Hermetic and alchemical thought).

and Conway. It is part of the *fortuna* of both Conway and Cavendish—a point to which I shall return later.

Another example of feminist scholarship of its time is Dena Goodman's groundbreaking work on the French *salonières*, which contrasted the intellectual leadership of women the *salons* of Paris with the Masonic lodges which were male-dominated and affording the repressive model into which *salons* eventually degenerated. As Margaret Jacob and Janet Burke have argued, this anti-masonic account derives from outdated, hostile views of earlier historians (Furet and through him, Cochin). Archival research, Jacob and Burke, tells a very different story of eighteenth-century freemasonry and the role of women within the ranks of the freemasons.[10]

These examples underline the fact that the historian's perspective is itself time-bound. And this applies to historians of philosophy as much as to feminists and other readers of the philosophy of the past. Contemporary context is not the only context that affects the reading and interpretation of philosophy. There is also the context of reception, of the philosophical cultures within which philosophy is received, be these in the past, in other countries and cultures, or at the present time. A fact which is often overlooked is that the relationship between the original historical context and later contexts, or contexts of reception, is an integral part of the history of philosophy. So, the gap between the past and the present which the mismatch of methodologies has opened up—the disconnection, if you like—is not an unresolvable problem. In fact, it need not be a problem, but an opportunity. To recognize the changes and differences between past and present opens the way for pursuing another line of enquiry, namely by means of tracing the fortune of philosophers and philosophies across time. The historical context is the ground base of *fortuna*, firstly because of its importance for the recovery of female philosophers, and secondly because this, in turn, opens the way for comparative context.

As I have argued elsewhere, the difference between philosophy as we read it today and that philosophy as it was received in the past are historically interesting for the questions that the alterity of past arguments raise.[11] These include questions about how a philosopher or philosophy acquired came to be read as they are today, how philosophies and the interpretation of philosophers have developed and transformed over time, and why this should be so. A key issue is the question of interpreting philosophy across time, and this is a dimension of the history of philosophy. To answer that question entails following the historical trajectory from the past to the present, examining the philosophical debates and the re-workings of philosophical ideas in new and different circumstances which occurred across time, tracking the standing of philosopher of a particular period with his or her standing among later generations. It requires, in other words, tracing a philosophical *fortuna*. Doing so allows us to confront not just matters of reception of ideas and arguments, but also the issues of inclusion and exclusion, amnesia and recollection specific to the fate of

[10] Burke and Jacob (1996).

[11] Hutton (2015).

female philosophers.[12] Eileen O'Neill made a start in tracing how we got from there to here in her now-classic essay, "Disappearing Ink. Women Philosophers and their Fate in History".[13] That essay took a retrospective view of the history of philosophy and laid out the stall for the recovery of forgotten women philosophers, or, rather, "threw down the gauntlet" to historians of philosophy, because it challenged them to address the neglect of women. But the history of philosophy is not just about the forgotten past; it comes right up to the present. A philosopher's *fortuna* is, therefore, something rather more modulated than simply her "fate", because tracing a philosophical *fortuna* entails following the historical trajectory from the past to the present, examining the philosophical debates and the re-workings of philosophical ideas in new and different circumstances which occurred across time right up to the present. Understanding how a philosopher or philosophy came to be interpreted (or not, as the case may be), is important for tracing influence and helps us to see how past philosophy connects with later periods and contributes to subsequent philosophical developments. In the case of women philosophers, to trace their fortunes offers a means to examine the impact of their thinking, the conditions in which it was received. Importantly, it also focuses on the positive as well as the negative, on how, why, and when they came *included* in the community of philosophers, not just explaining their *exclusion*. Even though inclusion means acknowledgement of their existence, it does not necessarily mean that they were accorded full recognition as philosophers. So, considering someone's *fortuna* also means thinking about whether they were treated as *philosophers*, how else they were viewed, and when and how they came to be considered philosophers. Thus, the question of a woman's identity as a philosopher is central to the *fortunae* of female philosophers. Focusing on philosophical *fortunae* has the potential to bring the history of women philosophers full circle with the present.

3.1 *Fortuna* and the Female Philosopher

Applied to women philosophers, the *fortuna* approach has its special challenges, arising from historical neglect of women by philosophers and historians alike. For the second part of this paper, I shall address those challenges, I shall then conclude by suggesting how we might use available sources to as evidence, by taking a lesson, so to speak, from Mary Hays.

Given that so many female philosophers have been consigned to oblivion, to trace the *fortunae* of philosophical women would seem to be an unpropitious way of going about writing their history. For a long time, it has seemed that women's contributions to philosophy went nowhere and the panorama of women's contribution to philosophy was a sorry picture of abortive endeavors. The "fate in history" of women philosophers identified by Eileen O'Neill was obliteration. Such is the disconnect

[12]cf. Ruth Edith Hagengruber's observation that, "The history of philosophy as a whole is … a record of inclusions and exclusions, of forgetting and rediscovering," Hagengruber (2015), p. 1.
[13]O'Neill (1998).

between past and present, between present women's philosophy and women philosophers of the past, that there appears to be no *fortuna* to trace. This would seem to indicate that there is no "history that lies between that philosophy and the present day", and in most cases there appears to be no "actual influence" (to quote Bernard Williams). On this scenario, much of the *fortuna* of women's philosophy is likely to yield a panorama of dead ends. But this perception is changing, as the work of recovery progresses because we are now discovering more evidence of the reception of women philosophers across time.

Of course, there is no question that there are women whose philosophical legacy has been lost or forgotten beyond retrieval for the absence of sources. But there are also others whose philosophy has simply been disregarded, whose books have lain unread on library shelves for centuries, only to be picked up at a later date. This is true for many male philosophers, as well as women: the history of philosophy is in fact full of such instances, for example, the recovery of Plotinus' philosophy and the revival of skepticism in the Renaissance. It is largely the case that many, perhaps most, women philosophers of the past were forgotten until, in the wake of the woman's movement of the late twentieth century, feminist philosophers set about unearthing them and identifying them. But some of them did not have to wait until modern times, because they were brought back to notice by historians or philosophers in earlier periods. However, it is not just the fact that they have restored to visibility that is relevant here. Also relevant is when this happened, and in what context, also the question of whether they were always credited with being philosophers. As soon as we ask those kinds of questions we begin to construct *fortunae* of these philosophers. So, for example, the evidence that they once existed often comes from earlier histories of philosophy, for example, Diogenes Laertius, and Thomas Stanley's *History of Philosophy* (1656). A particularly rich collection of names is to be found in Gilles Ménage's *Historia Mulierum philosopharum* (1690). Not all the texts which mention philosophical women were histories of philosophy—for example Marguerite Buffet's *Les Eloges des Illustres Sçavantes, tant Anciennes que Modernes* which is appended to her *Nouvelle observations sur la langue Françoise* (1668), or George Ballard's, *Memoirs of several ladies of Great Britain* (1752).[14] Many of the women who figure in these works are just names, with little or no detail of what they thought. But these histories do, at least, do them the credit of recognizing that they were philosophers. The question needs to be asked how they get to be included, and, of Ménage, why at that point in time in the seventeenth century did he come to include so many women philosophers? How many of them continue to be acknowledged in subsequent histories is another matter, but that too is the business of *fortuna*.

There are of course, also many cases of women thinkers who have not been forgotten, but whose philosophical credentials have dropped from view or been distorted beyond recognition. This usually occurs outside the context of a history of philosophy. Among these are truly scandalous cases where invented reputations have served to disqualify women of their entitlement to be considered philosophers, women who have been re-invented by later commentators as something that has nothing to do with

[14] Buffet (1668), Ballard (1752).

philosophy: Conway as an hysteric; Cavendish as an eccentric if not madwoman, Du Châtelet remembered as Voltaire's lover, and not as a great mind. But there are many other ways in which a woman's philosophical star might be dimmed. A number of women are memorialized not for their philosophy but for their scholarly learning (Marie Jars de Gournay and Princess Elisabeth, for example), or their piety (Mary Astell, for example). (George Ballard was at pains to stress the piety and virtue of the women featured in his *Memoirs*).

There are other cases where literary studies and feminist imperatives have ensured women like Mary Astell, Mary Wollstonecraft, and Margaret Cavendish, have not been forgotten, but, until fairly recently, their non-philosophical interests have obscured their *philosophical* ones. (To be fair, some of the responsibility rests with philosophers, who have been all too ready to accept that the genres in which these women wrote were non-philosophical). Madame de Scudéry, another woman who is remembered as a novelist, not a philosopher. She certainly was a hugely successful novelist, with an international readership, as attested by translations of her work into English, Spanish, Italian, German, and even Arabic. From early on, she was treated as a pedantic *précieuse,* attacked by Boileau; only recently claimed as a philosopher.[15]

The patterns of fragmentation and distortion, remembering and forgetting that has been the lot of most philosophical women are part of the *fortuna* of each. Even if most of these philosophers are known by name only, an important aspect of their *fortunae* when they were recognized as philosophers, and, in the case of those who have extant writings, how much of their arguments were discussed. When tracing the *fortunae* of women philosophers, what is more difficult to discover is whether was any take-up of their ideas, or engagement with their arguments. This is the case even those for whom philosophical writings are extant. Gaining a wider picture of women's contribution to the general philosophical conversation is hampered partly for the reason that many did not publish, or published anonymously. But also, because in the nature of philosophical debate, they don't circulate with authorial labels when they are absorbed into the general conversation of philosophical discourse (even if philosophical ideas and arguments come to be associated with particular philosophers). Some examples of philosophers recognizing their female peers are well known, of course—for example, Leibniz's comment in his *Nouveaux essais* that his philosophy is close to that of Anne Conway. There are also examples of discussions between male and female philosopher—Leibniz with Lady Masham, Norris with Mary Astell, Elisabeth with Descartes.[16] In the case of Princess Elisabeth, whose philosophical views have to be reconstructed from her correspondence with Descartes, one could argue that Descartes' *Passions de l'âme* is evidence of the take up of her ideas, and therefore part of her *fortuna*, since it was in response to her that Descartes composed the work. Evidence for women philosophers being part of the conversation is easier discerned among contemporaries, and through correspondences—not that there are many such sources. The impact of such sources is beginning to be actively explored.

[15]Conley (2011), see also Green (2014). Madame de Scudéry was partly rehabilitated by Victor Cousin in the mid-nineteenth-century.

[16]See Hutton (2014).

Again, so far this kind of investigation has largely been on women and contemporary philosophers.[17]

3.2 A Lesson from Mary Hays

In the last part of my paper I want to add substance to my claims that, despite the challenges, there is scope and material for reconstructing the *fortunae* of philosophical women. I shall do so by means of a concrete example: Mary Hays' *Female Biography or Memoirs of Illustrious and Celebrated Women of All Ages and Countries* (1803). Feminist and friend of Mary Wollstonecraft, Hays is another example of a woman who is largely unrecognized as a philosopher: she is normally classed variously as a historian, a feminist, a rational dissenter or a biographer rather than as a philosopher. The interest of Hays for present purposes is not her philosophy as such, but the fact that she affords scope and material for the fortunes of many early women philosophers. The primary aim of Hays's *Female Biography* was to give women the recognition they deserved and to correct the tendency of historical biographies to overlook them.[18] Hays' feminist motivations can thus be seen within the context of Mary Wollstonecraft's feminist "vindication" of women.

Hays's *Female Biography* includes entries for women from many walks of life, among them philosophers. In the grand scheme of her book, philosophers are a tiny minority, and the entries which discuss their philosophical views even fewer. Nevertheless, her *Memoirs* are striking for Hays's high valuation of female intellect. Her bright constellation of intellectual women includes many whom we would recognize as philosophers: Astell, Cavendish, Du Châtelet, Cockburn, Diotima, Elisabeth of Bohemia, Gournay, Hipparchia, Hypatia, and Masham. There are omissions: neither Conway nor Wollstonecraft figure in Hays's *Memoirs*.[19] Hays does not recognize all these women as philosophers: the long entry on "Madame de Jars" (i.e., Gournay), for example, focuses on her standing as a learned lady, providing a list of the learned men who consulted her. And Madeleine de Scudéry (another long entry) is presented as a novelist. Although Hays' judgment is by no means complimentary (her novels are pronounced tedious and prolix), she admires her "ingenuity" and "elevated sentiment". The entries on ancient philosophers are rather perfunctory. Diotyma (sic) rates a single sentence. The entries for Hypatia and Hipparchia are chiefly anecdotal. In many cases, Hays simply follows her sources, as in the case of Margaret Cavendish, which largely repeats Ballard. (It is notable that she is positive about Cavendish). The short entry on "Elizabeth Frederica of Bohemia" notes that Descartes praised her as

[17] See for example Broad (2006).

[18] Hays (1803). Modern edition 2013–14, ed. Anna M. Fitzer & Gina Luria Walker. On Hays, see Walker (2006), James (2012).

[19] Since Conway's book was published anonymously, her authorship was not picked up by the biographical anthologies on which Hays depended for her sources (e.g. Ballard and *Biblioteca feminaeum*). Hays did publish a separate appraisal and appreciation of Mary Wollstonecraft.

"the only person whom he had ever met that perfectly comprehended his works". The long entry on Queen Christina of Sweden notes her "taste for abstract speculations, for the severer sciences" and the esteem in which she was held by Descartes, Gassendi, and Bayle. The supreme case of female intellect, for Hays, is Madame Du Châtelet (Marchioness De Châtelet), notable for her "uncommon capacity and vigor of mind", who "may be said to have rivalled both Newton and Leibnitz" in mathematics. "Her works afford a proof of the power and force of her mind, and of the capacity of her sex for profound investigation and scientific research: she deservedly ranks among the first philosophical writers".

In many cases, Hays does little more than flag up women of philosophical bent. But there are instances where she summarizes, or even engages with their views. An example is her entry on Catharine Trotter Cockburn. Although much of this is devoted to Cockburn's conversion to and from Catholicism, a good deal of it is devoted to her philosophical interests: her moral philosophy, logic, and the strength of rational arguments are singled out for mention. "She took a peculiar pleasure and interest in metaphysical inquiries, so captivating to an acute and inquisitive mind…" (p. 158). Furthermore, Hays actively intervenes in the philosophical assessment of Cockburn, by comparing Cockburn's opponents to contemporary utilitarians who deny that virtue arises from innate moral sense; and defended Cockburn by invoking Clarke's idea of universal fitness of things.

Hays's book is important for several reasons: for preserving the memory of forgotten women, for its positive evaluation of their achievement, and for acknowledging their intellectual capacities. She ensured that among the different callings of womankind the philosophers were not wholly forgotten, that their being philosophers was regarded as a valuable achievement, and that the female intellect was something to cherish. In all these respects the bookmarks a positive moment in the fortunes of women philosophers, particularly those whose philosophical credentials have been underplayed, if not dismissed in other accounts (e.g., Madame Du Châtelet and Margaret Cavendish). Hays's philosophical engagement with her subjects may be modest, but it is not the less important for that. Their significance is first that Hays considered that philosophical evaluations should be included. Secondly, they give an indication of the reception and interpretation of the particular women discussed in the post-revolutionary period. Thirdly, Hays's book affords material from which we can make comparative judgments about the treatment of these women by other biographers (e.g., Ballard). Furthermore, the context of the book's publication is highly significant. A product of the Wollstonecraft agenda, it was a publishing success, which earned its author enough to support herself in her later life. That itself is a measure of the readership, and, by extension, the demands and interests of women readers. Its role in promoting women in nineteenth-century Britain remains to be explored. Hays's *Memoirs* is, therefore, a milestone in the fortunes of philosophical women.

References

Alanen, L. (2004). Descartes and Elisabeth: A philosophical dialogue? In L. Alanen & C. Witt (eds.), *Feminist Reflections on the History of Philosophy*. Dordrecht, Holland: Kluwer Academic Publishers.

Alanen, L., & Witt, C. (Eds.). (2004). *Feminist reflections on the history of philosophy*. Dordrecht: Kluwer Academic Publishers.

Alcoff, L. M., & Potter, E. (Eds.). (1993). *Feminist epistemologies*. London: Routledge.

Ballard, G. (1752). *Memoirs of several ladies of Great Britain, who have been celebrated for their writings or skill in the learned languages arts and sciences*. Oxford.

Bordo, S. (1987). *The flight to objectivity. Essays on cartesianism and culture*. Albany, NY: State University of New York Press.

Broad, J. (2002). *Women philosophers of the seventeenth century*. Cambridge: Cambridge University Press.

Broad, J. (2006). A Woman's Influence? John Locke and Damaris Masham on moral accountability. *Journal of the History of Ideas, 67*(3), 489–510.

Buffet, M. (1668). *Nouvelles observations sur la langue Françoise, Où est traitté des termes anciens & inusitez & du bel usage des mots nouveaux. avec les Eloges des Illustres Sçavantes, tant Anciennes que Modernes*. Paris.

Burke, J. M., & Jacob, M. C. (1996). French Freemasonry, women and feminist scholarship. *The Journal of Modern History, 68*(3), 513–549.

Code, L. (1982). *What can she know? Feminist theory and the construction of knowledge*. Ithaca: Cornell University Press.

Condren, C., Gaukroger, S., & Hunter, I. (Eds.). (2006). *The philosopher in early modern Europe: The nature of a contested identity*. Cambridge: Cambridge University Press.

Conley, J. M. (2011). Madeleine de Scudéry. In E. N. Zalta (ed.) *The Stanford Encyclopedia of Philosophy* (Summer Edition). Available from http://plato.stanford.edu/archives/sum2011/entries/madeleine-scudery/. Accessed 1 Dec 2016.

Dykeman, T. B. (Ed.). (1999). *The neglected canon: nine women philosophers*. Dordrecht and London: Kluwer.

Elisabeth of Bohemia. (2007). *The correspondence between Elisabeth, Princess of Bohemia and Descartes* (L. Shapiro, introduction and translation). Chicago: University of Chicago Press.

Garber, D., & Ayers, M. (Eds.). (1998). *The Cambridge history of seventeenth-century philosophy* (2nd ed.). Cambridge: Cambridge University Press.

Gatens, M. (1991). *Feminism and philosophy: perspectives on difference and equality*. Cambridge: Polity Press.

Gilligan, C. (1982). *In a different Voice: psychological theory and women's development*. Cambridge MA: Harvard University Press.

Green, K. (2014). *A history of women's political thought in Europe, 1700–1800*. Cambridge: CUP.

Harding, S. (1991). *Whose science. Whose knowledge? Thinking from Women's Lives*. Ithaca: Cornell University Press.

Hagengruber, R. (Ed.). (2012). *Émilie du Châtelet: between Leibniz and Newton*. Dordrecht: Springer.

Hagengruber, R. (2015). Cutting through the veil of ignorance: rewriting the history of philosophy. *The Monist, 98,* 34–42.

Hays, M. (1803). *Female Biography; or, memoirs of illustrious and celebrated women of all ages and countries*. London. Modern edition 2013–14, by Anna M. Fitzer, & G. Luria Walker London: Pickering & Chatto.

Hutton, S. (2004). *Anne Conway, a woman philosopher*. Cambridge: Cambridge University Press.

Hutton, S. (2014). Religion and sociability in the correspondence of Damaris Masham (1658–1708). In S. Apetrei & H. Smith (Eds.), *Women and religion* (pp. 117–130). Farnham: Ashgate.

Hutton, S. (2015). 'Blue-Eyed Philosophers Born on Wednesdays': An essay on women and history of philosophy. *The Monist, 98,* 7–20.

James, F. (2012). Writing female biography. Mary Hays and the life writing of religious dissent. In D. Cook, & A. Culley (eds.), *Women's Life Writing, 1700–1850: Gender, Genre and Authorship*. London: Palgrave Macmillan.

Laerke, M., Smith, J. E. H., & Schliesser, E. (eds.) (2013). *Philosophy and its history: Aims and methods in the study of early modern philosophy*. Oxford: Oxford University Press.

Lloyd, G. (1984). *The man of reason. 'Male' and 'Female' in western philosophy*. London: Methuen.

Merchant, C. (1980). *The death of nature: women, ecology and the scientific revolution*. San Francisco: Harper Row.

O'Neill, E. (1998). Disappearing ink: Early modern women philosophers and their fate in history. In J. A. Kourany (ed.), *Philosophy in a Feminist Voice. Critiques and Reconstructions*, (pp. 16–62). Princeton: Princeton University Press.

O'Neill, E. (2007). Methodological challenges to justifying the inclusion of specific women in our histories of philosophy: the case of Marie de Gournay. In E. Feder Kittay, & L. Martín Alcoff (eds.), *The Blackwell Guide to Feminist Philosophy*. Malden, MA and Oxford: Blackwell Publishing.

Popkin, R. H. (Ed.). (1999). *The Columbia history of western philosophy*. New York, NY: Columbia University Press.

Schiebinger, L. (1991). *The mind has no sex?*. Harvard: Harvard University Press.

Schmitt, C., Skinner, Q., & Kessler, E. (Eds.). (1988). *The Cambridge history of Renaissance philosophy*. Cambridge: Cambridge University Press.

Spender, D. (1982). *Women of ideas and what men have done to them: from Aphra Behn to Adrienne Rich*. London: Routledge & Kegan Paul.

Waithe, M. E. (ed.) (1987–1995). *A history of women philosophers*, 4 vols. Dordrecht, London and Boston: Martinus Nijoff/Kluwer Academic Publishers.

Williams, B. (2007). Descartes and the historiography of philosophy. In M. Burnyeat (ed.), *Sense of the past. Essays in the history of philosophy* (pp. 257–64). Princeton: Princeton University Press.

Williams, B. (1978). *Descartes: The project of pure enquiry*. London: Routledge.

Walker, G. L. (2006). *Mary Hays (1759–1843)*. Farnham: Ashgate.

Chapter 4
The Stolen History—Retrieving the History of Women Philosophers and its Methodical Implications

Ruth Edith Hagengruber

Abstract Women who are deprived of their histories can be compared to people who have lost their memories. They are unable to build a personal identity. This analogy may be the leading paradigm for my paper, which is primarily dedicated to epistemological questions. Throughout the last 40 years, many scholars have dedicated their endeavors to conserving the writings of women philosophers. Now we have access to valuable sources that show that the history of women philosophers stretches back as far as the history of philosophy itself. Using the history of women philosophers as a methodical approach to philosophy is a unique and indispensable means to widen and to change philosophical insights. Re-reading the history of philosophy and including the ideas of women philosophers, however, does *not* only add some more narratives but challenges the methodology of philosophy. The history of philosophy that we are traditionally educated in the western world is simply not true to the *facts*. Thus, I demand a rewriting of the history of philosophy that takes into account the ideas that are incorporated in the writings of women philosophers and that have been denied by the narratives and fabric of sexualized and patriarchally influenced thought.

4.1 Rewriting the History of Philosophy

Throughout the last 40 years, many scholars have dedicated their endeavors to conserving the writings of women philosophers. Today we have access to valuable sources that show that the history of women philosophers stretches back as far as the history of philosophy itself. Edith Stein, Hannah Arendt, and Simone de Beauvoir did not appear out of the blue.

Using the history of women philosophers as a methodic approach to philosophy is a unique and indispensable means to widen philosophical insights. The rise of

R. E. Hagengruber (✉)
Center for the History of Women Philosophers and Scientists, Paderborn University, Paderborn, Germany
e-mail: ruth.hagengruber@upb.de

S. Thorgeirsdottir and R. E. Hagengruber (eds.), *Methodological Reflections on Women's Contribution and Influence in the History of Philosophy*, Women in the History of Philosophy and Sciences 3, https://doi.org/10.1007/978-3-030-44421-1_4

women philosophers' participation in philosophy is connected to the re-reading of its history. Re-reading the history of philosophy, however, challenges the methodology of current practice. The exclusion of women philosophers from the history of philosophy is a result of a centuries-old practice of the gender-minded self-interest of certain groups, supported by a culturally established patriarchal hierarchy. After centuries of exclusion, the perspective is now changing.

In 1987, Mary Ellen Waithe presented the first *History of women philosophers* written three hundred years after the one from Ménage (1690/1984). Waithe provided a collection of information on women philosophers from Antiquity up to the present in four volumes, mainly taken from the western philosophical tradition (1987–1995). In her introduction, she explains that she had never heard of any women philosophers who had lived prior to the twentieth century, with the exception of Hildegard of Bingen and Queen Kristina of Sweden (1987, ix). Some years before, Beatrice Zedler, a contributor to Mary Ellen Waithe's edition, had translated the *Historia Mulierum Philosopharum*, a collection of sources on 67 women from Antiquity, first published in the book by Ménage (1690/1984). These sources, as well as much following information, have challenged scholars of the history of philosophy to take a new direction in its interpretation.

Since then, a large amount of research has been done on reconstructing the ideas of women philosophers. Many scholars have dedicated their endeavors to conserving their writings, so we now have access to valuable text sources. We have gained access to the ideas of ancient women: the Pythagorean women play a dominant role in rediscovering texts from Antiquity (Pomeroy 1984, 2002, 2013; Allen 1997a; Theano 2010; Lazaroiu 2013); we are reevaluating the speeches of Diotima, and Aspasia, quoted in Plato (Stegmann 1910; Annas 1976; Tuana 1994; Haskins 2005; O'Mahony 2010; Blair 2012). The philosopher and astronomer Hypatia, who was famous in her epoch and widely recalled in Renaissance writings has regained attention (Dzielska 1995; Deakin 2007); Makrina, a sister of the Cappadocian teacher in early Christianity, reevaluated in the seventeenth century, has been rediscovered (Gregory1989; Frank 2000). The great Hildegard of Bingen's fame had never vanished, although she was not included in the philosophical canon (1997; Allen 1997b; Kerby-Fulton 2010; Lerius 2017). The eminent reputation Christine de Pizan enjoyed in her time was renewed in the twentieth century (1999; Willard 1990; Dascal 2005; Langdon 2002; Fonte 1997). Marie de Gournay was famous in her lifetime (2002; Ilsley 1963; Rauschenbach 2000); this is true also for Margaret Cavendish who inspired even Virginia Wolff's famous dictum of "a world, i.e., a room of one's own" (2001, 2003, 2004, Rees 2003; Whitaker 2003). Not to forget the French Salonières, Marquise de Sablé, the Marquise de Lambert, Mme Deshoulieres, women trained in stoic and skeptical thinking, who embodied the intellectual female figure of the seventeenth century (Harth 1992; Conley 2002). Anna Maria van Schurman (De Baar 1996), connected to the famous Marie de Gournay and best friend of Elisabeth of Bohemia, served as a role model a century later for Luise Gottsched, Emilie Du Châtelet, Laura Bassi and many more (Brucker 1745; Hagengruber 2018). Elisabeth of Bohemia, a Cartesian scholar, was as a role model herself for the intellectual women in the seventeenth and eighteenth centuries, and is re-gaining her fame through the endeavors of outstanding scholars (2007; Marshall 1998; Shapiro 1999; Bos 2010; McWeeny 2011). The influence of Anne Conway can hardly be overlooked (1992; Merchant 1979; Hutton 2004, 2012); Astell (1986, 2013; Perry 1986; Springborg 1995; Weiss 2004), Damaris Masham (2004, Frankel 1989, 1991; Hutton 1993) and Catherine Trotter Cockburn (Ready 2002; Sheridan 2007) are broadly discussed to reevaluate the philosophy of John Locke. Called the German *Sappho*, Luise Gottsched's sharp pen is receiving attention after 200 years (Goodman 1999, 2009). Emilie Du Châtelet tackled Newton, Leibniz, Wolff, Locke, Voltaire, La Mettrie and Mandeville and is honored today for her contributions to the renewal of philosophy and science (2007, 2011, 2015, 2018; Zinnser and Candler Hayes 2006, Kölving and Courcelle 2008; Hagengruber 2011; Hagengruber and Hecht 2019). Olympe de Gouges is known as a fighter for equal rights (1979, 2013; Schröder 1995; Muller and Bocquet 2012; Sherman 2013). Mary Wollstonecraft is an important figure in early feminist theory (1989; Falco 1996; Todd 2000; Craciun 2002; Karafili Steiner 2014), Sophie Germain (Klens 1996; Ornes 2009) belongs to the great women in mathematics and philosophy of the period, as well as Laura Bassi (Ceranski 1996) and Maria Gaetana Agnesi (Mazzotti 2007). Harriett Taylor-Mill (Mill and Mill 1970, 1998; Jacobs 2008; Bodkin 1999; Szapuová 2006), Charlotte Perkins Gilman (1975; Prasch and Sheth 1996) have become central figures for women's ideas in economics in the nineteenth century. Christine Ladd-Franklin (Furumoto 1994), Mary Calkins (2010; Madigan and O'Hara 1992; McDonald 2007), Susan Stebbing (Chapman 2013; Beaney and Chapman 2017), Constance Naden (1890; Alrabi 2012), E.E.C. Jones (Waithe IV; 25–49), Victoria Lady Welby (Peirce and Welby 2001; Petrilly 2009) are testimonies for women's contributions in analytic and pragmatic philosophy (Rogers 2005). Alexandra Kollontai was an eminent figure of Russian philosophy in the early twentieth century (Lokaneeta 2001; Farnsworth 2010; Porter 2014). The contributions of Edith Stein (1989; 1990–2014; 2006; Calcagno 2007; Lebech 2015), Hedwig Conrad Martius (Bello 2002) and Gerda Walther (McAlister 1992; Calcagno 2018) extend our perspective of twentieth-century phenomenology (Luft and Hagengruber 2018). Although Ayn Rand is controversially discussed, she is a widely read author (Rand 1961; Reisel Gladstein and Sciabarra 2003; Britting 2004). Hannah Arendt and Simone de Beauvoir enjoy a broad reception today and many books and scholarly investigations have been dedicated to their ideas, though for too long the public interest reduced their work as having grown out of a dependency on their male companions, Heidegger and Sartre (Villa 1995; Ettinger 1997; Meier-Katkin 2010; Daigle and Golomb 2009).

4.2 Cornerstones of Another History of Philosophy

If the records stand as shown, why then is the history of philosophy as presented in the canonical textbooks still the history of the ideas of men? We need to rethink the history of philosophy in the face of this huge amount of material. The history of women philosophers presents a philosophical testimony and legacy from 2600 years of European philosophy, and this evident proof of its writings requires a response to the question of why it is still propagated that women's ideas did not contribute to the modeling of the western canon. Did they not or is it not known that they did? Was this kind of habitude the same in all periods? The above-mentioned multitude of texts written by women philosophers indicates that there were epochs when women's voices were heard. They give reason to wonder whether the twentieth century, often thought of as the century of women's breakthroughs, was long and dark with regard to the fact of women's philosophical participation.

The testimonies from Antiquity present women as ideals of philosophical striving but also as philosophical speakers. In Parmenides' writing, the women assign the way to philosophy.[1] They are still role models in Plato's *Symposium*. Socrates, the most celebrated philosopher in western history, famous for his philosophical renewal and his challenge of religion and culture was trained by two women, Diotima and Aspasia, as Plato reports. Socratic shimmering sexual identity is part of the Socrates figure, who relates this methodology to his mother's art of maieutic. His teacher in rhetoric, Aspasia, was charged with the death penalty for seducing the youth and for blasphemous speech, and so was Socrates. He suffered the same accusations as Aspasia, and this refers to, as Aristophanes tells us, the *sophistra acutissima* (Coventry 1997). Plato's reference to female practice is repeated in the speech delivered by Diotima and presented in the *Symposium*, and was a piece of hard training for the young Socrates, who was on the wrong path, idolatrizing the objects of beauty instead of understanding that the good and the beautiful must be created, as Diotima tells him when she teaches him the way to eternity and beauty by "birthing" the concepts in the middle between good and bad. In his *Republic,* Plato's Socrates challenges the view of his audience and provocatively demands equal education for everyone (including girls) and the participation of women in the philosophical class of guardian kings. To separate women from any kind of hierarchy, he demands dissolving family ties. This was often heavily criticized by male and female interpreters. As far as we are concerned, asking how history evolved, what came with it and what was forgotten and neglected, it is important to know that also Platonic philosophy was widely forgotten and even suppressed for hundreds of years. Hypatia's murder in Alexandria, her lost fight, and the emerging Christian culture sealed the downfall of women's education for centuries before it was reintroduced and celebrated in the Renaissance epoch. However, the twentieth century was also very critical of Plato's concern for women for a long period (Annas 1976; Wider 1986; Irigaray 1992; Bar On 1994; Cavavero 1995). Hypatia's murder by the Christians was a symbol of the decay of Platonic Academism, which was also open to women. Inaugurated by Plato, it was practiced in the tradition of his academy. Thus, Platonic philosophy fits the exclusion and recovery scheme, as does the history of women philosophers. And indeed, the history of women philosophers parallels the rise and fall of Platonism. After having been broadly sanctioned for a long period, its reentry into European culture became a crucial catalyst in the Renaissance epoch (Schmitt et al. 1988).

[1] See Songe-Moeller's contribution in this volume. See also (Acker 2013).

Although so much has been written about the Renaissance phenomenon in our culture and though everyone is familiar with the idea that the entry to the world of scientific and methodic reasoning began in the Renaissance, few, with the exception of Margret King, have pointed out that many outstanding intellectual women have been involved in this cultural rise and the rediscovery of ancient and pagan philosophical writings and ideas. When Platonist and ancient writings pushed the world toward modernity and scientific interest in Europe, the perspective on women also changed, as the documents prove.

With regard to his presentation of science and his view toward women, Aristotle was consistent with the scientific and religious perceptions of Christian and Muslim theology and philosophy. He held that the earth was the center of the universe and he thought women to be the container for male seed. Aristotle was the philosopher of observation and this was what he had observed. Platonic philosophy—in its textual source, not in its neoplatonic interpretation—pled for the contrary: Not to believe what you see and to question whatever you observe, as everything you see is deceiving. Men and women are chained in the cave of shadows. The Platonic dialectic methodology, as seen in the *Republic* and his *Symposium*, presented women who questioned what was considered reality. Many texts from Antiquity came to Italy for the first time and changed the universe of conceptual hierarchies. The Renaissance was a period that deserves its name exactly for its *re-reading* and *re-thinking* of ideas. This *re-reading and re-thinking of history* is at the core and is the source of early modern philosophy. It is its essence and the core of its methodic renewal.

While the return of Antiquity has been widely discussed, the very important fact that herewith the status of women was also turned upside down has been simply ignored and one may soon find out why. Outstanding scholars, such as Schmitt et al. (1988) contributed to this history of the Renaissance but neglected to note that the new forms of philosophy and science were accompanied by a new perspective on and of the female. Women were represented differently. Think of the Madonna in churches from the Middle Ages, shown as the suffering mother. She re-emerges in the Renaissance as Primavera, Aphrodite, and Minerva by the very provocative Boticelli. The return to Antiquity implies the upgrade of women to ancient goddesses, to *Diana* and *Minerva* (Hagengruber and Rodrigues 2010). The celebration of these female goddesses signifies also, and at least partly, the reentrance of women into the public intellectual sphere. These societal changes came along with a substantial critique of the religious patriarchal institutions of the time.

Renaissance texts provide many contextual studies with references to the female "Platonic" context (Reeser 2016). Plato's mother and sister and the female students of the Platonic academy are now remembered by scholars of the period (Nails 2002). Hypatia is a profoundly established figure in the Renaissance perception of the Antiquity, said to have been idealized by Raphael in his self-portrait in the famous *School of Athens* in the Vatican *stanze*. Beyond the new vision and revaluation of women in the paintings of Botticelli and Raphael, women also entered the scientific sphere. Hypatia was seen as having inspired the heliocentric cosmology of Copernicus, being a member of the Platonic tradition and critical of geocentric cosmology (Grafton 1997).[2] During the Renaissance, Platonic academies were re-established, some of which had female leaders (Yates 1983). The narrow academic framing of Catholic universities, devoted to raising clerics, was partly overcome. The university of Bologna, emerged as a secular university earlier, included women from time to time; most famously when Laura Bassi became a professor there and occasionally this tradition of including women professors was remembered (Ceranski 1996, 51, Hagengruber 2018).

It is well known that Platonic writings regained huge importance in the Renaissance period. However, Renaissance Platonism was taught through the Neoplatonist take of the Florentine philosopher, Marsilio Ficino. He was praised for the edition of Platonic writings in the second half of the fifteenth century and for his idealization of beauty beyond the reach of men. Scholarly reference to Ficino is one-sided and overlooks the huge debate on the interpretation of Plato, to which women vividly contributed. Intellectual women wrote on Plato's *Symposium* and gave very different interpretations to those Ficino provided. While Ficino perpetuated a Plato interpretation of idealism beyond the real, this construction of the Platonic text and Diotma's speech in the *Symposium*, were turned the other way around by Tullia d'Aragona, who made the case for the dignity of physical love. Ficino's fleshless metaphysical interpretation, putting love and the right and the good in a world beyond ours, idealized what humans strive for. He also "idealized" Diotima. As Mary Ellen Waithe found, to her knowledge no one before had ever claimed that this women did not exist (Waithe, *vol.* I, 106). The concept of love, as presented by Diotima, inspired many intellectuals in this period. Tullia d'Aragona's *Dialogue on the infinity of love* is part of a prolific series of essays on love and served as a model for many of the female philosophers to come in the next centuries. Lucrezia Marinella's essay from 1591 on *The Nobility and Excellence of Women* refers explicitly to Aspasia and Diotima and claims the superiority of women. She was neither the first nor the last to refer to them. The Platonic females became a stable part of European culture, and countlessly remembered foremothers for Marinella, Marie de Gournay, Emilie Du Châtelet, Laura Bassi and many others. However, their legacy was meaningfully divided. While Diotima, said to have been a priestess, was idealized, Aspasia, the powerful wife of Perikles, was defamed throughout tradition and history although she was no less influential in western culture. Diotima and Aspasia become the most famous but also the most critically and controversially discussed representatives in male-stream—mainstream as well as in feminist research in the twentieth century (Neumann 1965; Irigaray 1992; Cavavero 1995; Nye, 1989; Allen 1997a; Henry 1995; Salkiever 1993; Ciempiel 2008; Stegmann 2010; Asimopulos 2011; Beiser 2012).

[2]"Several major Greek works, above all the Almagest owe their survival to her. Her commentary on the works of Apollonius was also important. Since Copernicus could not have worked without the Almagest, Hypatia must be regarded as a significant contributor to modern astronomy", Grafton 1997, 261.

The Renaissance retrieval of the philosophical women of the Antiquity, of Sappho, Diotima, the Pythagoreans, and so forth, found its peak in the 1690 publication of Gilles Ménage's sourcebook on women from Antiquity, the *Historia mulierum philosopharum*. The question Ménage puts at the beginning of his book to take position in this disputed field of scholarly study is telling even today: "The number of women writers is so great that it is even more astonishing that some of them could have been believed to have been single phenomena, or exceptions to the rule" (Ménage 1690/1984, 3).

This proves that these contributions to re-establishing the memory of women's rich tradition have not been singular either but were part of a long history which presented itself from various perspectives. From the late Middle Ages on, lists were created to demonstrate the influential history of intellectual women, tracing their traditions back into mythical ages. The *City of the Ladies* written by Christine de Pizan follows this model. Boccaccio raised the awareness of a long tradition of women in his *De claribus mulieribus,* as he did in his *Genealogia deorum*, a collection of female figures from Livius, Ovid, Plinius, Sueton, Tacitus, Servius, and Lactantius (King 1991). Germany provided a rich tradition of such lists, naming many women from "babylonic times" up to the present, from the famous *Frauenlob* in 1631 to *Paullini* and *Eberti* in 1705/06 and *Brucker* (Hagengruber 2018).

Hand in hand with intellectual insights, women demanded societal changes. How much the world of that epoch was shattered by women's demands is evident when considering the very famous utopias of that time, hitherto seen as inspired by Platonic ideas. Thomas Morus' *Utopia* and Tommaso Campanella's *Sun City* provide two elaborate though different approaches of why and how women are equal, and why they consequently have to share work as well as societal and political responsibility. Two different models are presented that give a glimpse of the discussion in this epoch. While Morus follows a strictly egalitarian program, Campanella organizes a model of mutual accomplishment, in which the tasks to be accomplished by women differed from those of males, this being the best way to flourish as it demanded the best from everyone. It was Campanella who called his age the "epoch of the reign of women", "receiving now the rights, they have been deprived of for so long" (Hagengruber 2015, 151).

The intensive absorption with ancient philosophy was not confined to Italy in the fifteenth and sixteenth centuries. It also passed to France, Germany, and England. Gilles Ménage's *Historia mulierum philosopharum* was published and republished in 1692, in English in 1702, in French in 1758, reprinted in 1761, and Latin available in 1830, before it was "rediscovered" in 1984. Also in the eighteenth century, the German cultural magnate, Christoph Martin Wieland, with his mentor and mentee, Duchess Anna Amalia of Brunswig-Wolffenbüttel and Sophie La Roche (Strohmeyr 2006), was an influential writer and journalist, who translated and published the texts from Pythagorean women philosophers for his *Historischer Calender für Damen des Jahres 1789* and wrote his *Defense of Aspasia* (Wieland 1796).

4.3 Criteria of a Newly Organized Canon

In light of information gathered within the last decades, we are beginning to think about how the story of the history of philosophy will become different from the one we are familiar with and which has been written down. What we easily learn from the new material is the fact that in certain epochs women played an active part in philosophical debates, and their writings have been known, disputed or have even remained a part of the debates throughout the ages. The education of the present, however, has relied on histories of philosophy which took no notice of women's ideas, partly ignoring and devaluating them. In part, we did not know that certain ideas had been drafted by women philosophers, as women's ideas were also appropriated into the canon by acts of outright plagiarism, as Waithe reports for example of E.E.C. Jones and Bertrand Russell (Waithe and Cicero, IV, 25–49).

In view of this huge amount of writing, we understand that there was intentional and active exclusion, which ostracized women's ideas from school canons and universities. This is different from the traditional narrative excusing the absence of women; that they're being missed was a consequence of women's lack of capacity or their forbidden access to education or materials. Restricted access was not the primary cause, but a secondary reason to explain ostracism. Testimonies prove that parents, societies, academies, and philosophers tried hard to bring women in at all times. In his *History of the most famous writers* Brucker holds there were women in all nations and at all times, who were as gifted as men, but were restricted by societal order (Hagengruber 2018). A history of women's intellectual courage and fortitude needs to be written, honoring all those who fought against the odds of a dominant cultural and patriarchal pressure. Aspasia was sentenced to death, Hypatia was murdered for the influential position she held in Alexandria, Olympe de Gouges was led to the guillotine because she attacked the bloodthirsty Robespierre, to name only the most famous among many feminine heroines. There is no reason to call each of them a "singular" phenomenon. They are part of a long tradition that was born in the moment that cultural power began to suppress women's demands for equality and freedom.

Thus, research into the history of women philosophers is more than merely historical research. It is a moral demand to rewrite the history, to uncover the omission and misuse women's ideas have been subjected to. This undertaking will lead to a new kind of philosophical tradition with new content. It will shape new categories and new topics as an important step in a critical revision of content and methodology to reveal implicit interests and self-advantage at the cost of women. This research is not the only means to renew philosophy, but it is an important means to free it from corrupt ideas propagated in self-interest. The reading of the history of philosophy is a methodical approach, as it is to read the *great books* anew. Those who reject this proposal, calling it impossible, because history is written in stone, should look at the stunning example of how Plato was rediscovered. Re-reading the history of women philosophers is a contribution to enlightening philosophy itself, and consequently, to criticizing the methodic underpinnings of an established history of philosophy that needs consequent and ongoing renewal.

4.4 Not a Gender Narrative: The History of Women Philosophers

The history of women philosophers stretches back as far as the history of philosophy itself. Edith Stein, Hannah Arendt, and Simone de Beauvoir did not appear from nowhere. Their work was based on their predecessors. They were standing, so to speak, on the shoulders of female titans before them. However, the awareness of women's own history and how it has been received was also quite different among the female representatives. Hannah Arendt was for example devoted to the history and conceptualized her ideas of a woman and a pariah with Rachel Varnhagen, an eighteenth century intellectual, called by Arendt her "closest friend, though she has been dead for some hundred years" (Weissberg 2000, 3). In contrast, Simone de Beauvoir, who quoted the names of many important female writers of the past in her essay on the *Second Sex*, essentially negated that feminists had contributed to history differently and by themselves, holding that women "mirrored the values of men" (Beauvoir 2011, 73).

Hannah Arendt never became famous for being a feminist thinker, although her writings show, as I hold, an awareness of the problem and imply a critique of women's exclusion. Simone de Beauvoir denied an active and alternative tradition of women thinkers before herself.

Judgements on the relevance of women in the history of philosophy has thus also been at stake in the discussion women provided on the topic with their interpretations of how to judge the absence of women in the history of philosophy. Since the great foremothers started to enter the public scene of philosophical discourse, the question of whether to interpret the universal attribute of doing philosophy in a gendered or non-gendered mode also evolved. Although rewriting the history of philosophy and including women philosophers is *not* one more narrative in the writing of history, it is true that inclusion of a gender history will add a gender narrative to the history of philosophy and also widen the view of this history. However, the point I want to make is the fact that the history of philosophy we are traditionally educated in the western world is simply not true to the *facts*. Thus, I do not ask for a widening of the perspective. I demand a rewriting of the history of philosophy that takes into account the historical facts that have been denied by the narratives and fabric of sexualized and patriarchally influenced thought.

Considering these texts and ideas that have been developed from Antiquity up to the present time, and which are in constant dialogue in the ongoing development of philosophy, thus *also* forming a part of its history, is not only an aspect which has been neglected up to now and which could be added by simple means. The absence of this history is a *fact* in the sense that it is a fact that only male children were chosen to be crowned as kings according to certain laws and that only certain families could produce the heirs to the crown. The exclusion of women was a *constitutive* decision of a cultural environment, which assumedly expected to thrive more successfully if it excluded women's participation in the intellectual, political and societal world. Rediscovering that nonetheless there has been activity by women in the relevant fields does add many *facts* to the history which has not yet been taken into account. Above all, it demonstrates that the exclusion of women as non-thinking and non-educated creatures can not be justified by a cultural narrative that is often repeated. The cultural narrative was incapable of suppressing the fact that women contributed to philosophy with vivid philosophical imagination, despite the hardest persecution and oppression. The exclusion of women's ideas from philosophical history has contributed to a badly performed scientific undertaking. *It is, as if we had calculated the planetary system of our sun, not taking into account every second planet.* These narratives of the incompetent women have doubtlessly dominated cultural development over centuries, but they could not destroy the vivid facts of ideas, thought by thinking women throughout history. The search for truth beyond the narratives and prejudices is driven by a strong will to *unveil the facts*. Not to possess all facts possible is not the same as justifying injustice as a narrative. We never know it all. The pure existence of women's ideas does not follow any narrative. These facts are not presented to glue a different perspective on the old history. These facts call for a different understanding of the world that will shatter our societies in many ways. To take into account the history of facts and the ideas of women, shows, "how historical forms are fabricated" (Munslow 2012, 79) and it forms an important contribution against violence and suppression.

Bringing in women's writing is a source of power for changing the narratives and therefore changing the future of our societies. It unveils another part of our reality; it is not only another *version* of telling the history of our world heritage. The part of our intellectual history involving women's ideas has been stolen. It has been taken away intentionally, it has been an activity that has at times been stronger and at times weaker. From time to time women were a part of this history, and thus their part in it must have been willfully removed. It is also important to understand that for the flourishing of societies compensation for this violent act is an important step to improve philosophy and its recording methods. The integration of women's ideas in a great history of philosophical ideas on freedom, morals, science, and knowledge is an important step to its improvement; it is a catharsis for immoral actions, a new start into a better future, as any abolition of immoral practices can be.

Rewriting the history of philosophy and including the intellectual contributions of women philosophers does not destroy the idea of a history of philosophy. On the contrary, it is a vivid demonstration of how important the history of philosophy is to philosophy itself as the basis of its own moral development.

4.5 Deconstructing Narratives

In accordance with the rich history of women philosophers, it is necessary to rethink the concepts and interpretations of those who have been missed in this history of philosophy and to question our perceptions and those of our culture to explain this factual absence. The exclusion of women, as well as the narratives which legitimated this, belong to an interpretational topology which has hindered the acknowledgement of their relevance to our cultural rise.

The exclusionary method functioned to establish and maintain the patriarchally formed canon and its outstanding achievements. Interestingly, the feminist analyses, for example in the work of Beauvoir, started with this explanatory narrative of the absent women to counter the misogynism in philosophy.

The absence of women philosophers in the history of philosophy formed the precondition for feminist theorists. The women found themselves bare of any tradition and continuity. Interpretational patterns formed by males to explain this absence came from anthropology and also from psychoanalysis. Women became the gender "without a voice" (Smith 1992, 2007). Invisible, and with no access to *language* and *symbols* to mark their own history, they explained the absence as a kind of biological and anthropological given. The essence of women, according to the representatives of French difference feminism such as Cixous, Kristeva, and Irigaray, is in being beyond the symbols of power, a mute companion in history. Women within this explanatory frame did not appear as intellectual identities, as they were the *other* to symbol, language, presentation, and power. The narratives of the speechless women confirmed a perspective on women's cultural presence which was compatible with what could be observed and it acted at an ontological level of the deprivation of intellectual participation.

The omission of women's ideas from intellectual history was also the starting point for Sandra Harding's critique of the epistemic framing of our traditional history. Harding focused on the factual absence of women, derived from their institutional erasure. Harding herewith confirmed the delivered history and built her feminist epistemology on the precondition that there were no women in the history of philosophy. From this premise, she concludes that the history of philosophy is not a relevant history for women philosophers at all, as women should not contribute to a tradition they had not been part of (Harding 1991). She built her theory on the cultural narrative of a history bare of women. Lloyd committed reason to the traditional way of doing philosophy and Harding reduced women's demands to a *standpoint philosophy* (Harding 1991, Lloyd 1984).

To a certain extent, feminist philosophy built its theory on a narrative of the absent female philosophers based on the patriarchally affected interpretations. That the history was not void of women but was intentionally denied to women, becomes evident in Beauvoir's work. When she drafted her thousand-page book *The Second Sex* (1949), she referred to an impressive number of women's names in the history of philosophy. Aspasia, Marie de Gournay, Emilie Du Châtelet, Mary Wollstonecraft and many more were mentioned, some of whom are now considered to form the canon of women philosophers and to which Beauvoir evidently had access while preparing in the libraries of Paris. Beauvoir, however, remarkably interpreted this as a womenless history, based on the fact that "this world has always belonged to males" (Beauvoir 2011, 76). Seemingly, she also denied credibility to the evident facts. The groundbreaking importance of Beauvoir's contribution led to an idolizsation of her interpretation. Even 40 years later, the famous historian of philosophy, Gerda Lerner, used Beauvoir's interpretative scheme to strengthen evidence that many women philosophers in the Renaissance were not of importance. Lerner ignored the ideas of Isotta Nogarola, Cassandra Fedele, Lucrezia Marinella, Christine de Pizan, Maria van Schurman and many more, repeating that "current scholarship holds that women made no significant impact on the writing of history until the late eighteenth century" (Lerner 1993/2013, 305). Feminist philosophers continued to hold on to this interpretational scheme, which is bare of women's ideas. Although Lerner was impressed by the large amount of historical material she had discovered, she did not consider these texts to be valuable or important to the history of philosophy. Lerner explained their minor value by acknowledging the commonly used exclusionary explanatory pattern, holding that earlier women did not possess the innovative potential to become outstanding contributors to the history of philosophy as "women were then educationally disadvantaged", and "did not significantly participate in the creation of the [culture's] symbol system" (Lerner 1993/2013, 304). This explanatory pattern was further employed in Margaret L. King's interpretation of women in the Renaissance. King successfully worked in the area of the history of women philosophers and started the outstanding University of Chicago Press series *The Other Voice in Early Modern Europe*, in which she publicized important material drawn from the Renaissance period. Although the interpretation of these texts remained in the above-explained pattern, we encounter the need to integrate women's texts into the history of philosophy in order to make up for the devaluation and as a sort of reparation: "The learned women of the Renaissance … received no degrees" and "exerted no great influence on emerging trends in the history of ideas," King claimed in her famous book on women in humanism and the Renaissance (2005, 81).

This kind of separate interpretation of feminist interest and historic analyses is confirmed in a report by Linda Lopez McAlister, which clearly shows the difficulties and challenges of research into the history of women philosophers. McAlister remarks that the approach to female history was filtered during that time. She calls this a kind of *litmus test* of feminist identity, separating feminist from non-feminist writers in the history of philosophy (McAlister 1989, 3). Again, creating and applying an interpretational category that branded research on the history of philosophy and enabled feminism to brand itself as a completely new concept was politics at the expense of women philosophers in history. From that perspective, women's philosophical contributions were expected to entail *different* topics than those of main-and male-stream writers. A tendency to classify women's writings as philosophically relevant only if they targeted specific and particular gender issues and if they testified to a suppressed and ostrasized history was established. H.L. Smith asks why feminist historians have shown little interest in women's contributions in the history of philosophy and named it the "paradigmatic separation" (2007; Green 2014, 5).

Thus, feminists have thereby confirmed the absence of women's intellectual contributions in history and have continued to spread this narrative themselves. However, the feminist critique in the male framed history of philosophy was outstandingly critical and exceptionally productive. For a long time, feminist philosophers may not have rediscovered women in the history of philosophy, but they have reviewed the history of male philosophers with great success and huge social impact (Stopczyk 1980). Although they (wrongly) accepted the absence of important female writers for a long period, and thus ignored the reconstruction of their own tradition, feminist philosophers have enabled a critical rewriting of the male-stream history, and thereby contributed to a critical rewriting of the history of philosophy itself by identifying misogynist arguments and strategies. Feminist philosophers succeeded in uncovering the ostracizing traditions in traditional philosophy and its canon, designed, not for the sake of philosophy, but for the sake of men's own achievements. "Feminist re-reading of the history of western philosophy was motivated initially by the desire to critically analyze philosophical misogyny" like Schott explains the feminist readings of Plato, Aristotle, St. Augustine, Thomas Aquinas, Spinoza, Rousseau, Kant, and Hegel (Schott 2007, 53, 56). For a long time, feminist philosophers did not reconstruct their own history but deconstructed the male-dominated history of philosophy. An example of this was delivered in the *Stanford Encyclopedia*, before it was essentially reworked in the Spring of 2015. Up to then, the Stanford article *History of Feminist Philosophers* provided only seven contributions dedicated to acknowledged feminist women philosophers in history, whereas 28 out of 35 topics referred to the reconstruction of the ideas of male philosophers.[3] This layout was changed when the growing interest in the history of women philosophers had brought the public to a more precise awareness of women in the history of philosophy. At that time a sizeable bibliography of texts of women philosophers throughout history was also added. On January 1, 2015, *The Monist* was dedicated completely to the presentation of *The History of Women's Ideas*. This was remarkable, as it was the first mainstream journal in philosophy that dedicated a whole issue to this neglected part of philosophy (Green and Hagengruber 2015). Since that time a change in the public reception has occurred. It is being acknowledged by a larger and broader philosophical audience that there is a history of women philosophers in their own right. In 2018, the Deutsche Gesellschaft für Philosophie established a permanent working group for the research of women in the history of philosophy. The interpretational pattern is now shifting from the deconstructive analysis of misogynism in the history of philosophy to a reconstruction of the history of philosophy to include women philosophers as a crucial method of writing the history of philosophy.

---

[3]This article was accessable till September 2014: plato.stanford.eduentriesfeminism-femhistbib.html#Gen (extracted September 10, 2014). It was substantially changed and presented a considerable bibliography on women in the history of philosophy in 2015, see now (Witt and Shapiro 2019).

4.6 Universal Claims Versus Gendered Interests

For quite a period it seemed that the research of women philosophers on the history of philosophy represented a specialized and gendered philosophical claim, meaningful to providing a narrative invention of gender history (Munslow 2012, 80). This separated women's contributions from the universal claim of philosophy. Following this philosophical attempt, philosophy became a place of specified and particularized interests. In considering women's perspectives, one had to ask how women with defined "gendered" and particular interests could have contributed to philosophy's universally conceptalized development. On the other hand, could men with sexist claims that demonstrate ostracism be taken as the rightful defenders of reason? That a group of feminist writers credited the male history as a history of reason and universalist values, while women installed themselves as the defenders of emotion and particular interests, followed the same pattern and was only the other side of the coin.

How could philosophers of any gender ever accept that a history that presented its interest in scandalous devaluation of women, as you can read in Rousseau, Kant, Fichte and many more, meet the universalistic demands of reason? How could this ever be recognized as the result of universal categories? It is, on the contrary, philosophy driven by misogyny and a disguised self-interest that only could work as effectively as it did because it was supported and protected by a culturally established patriarchal hierarchy. From Aristotle to Augustine and Nietzsche, we face misogynist texts that fail as philosophical texts presenting universal claims (Stopczyk 1980, Schott 2007). Genuine universal philosophical claims, construed to secure an unprejudiced critical approach, were rejected. The more texts of women philosophers printed, distributed and read, and so becoming part of our everyday cultural heritage, the better we will understand the grave philosophical limitations and mistakes the institutionalized mainstream and male-stream patriarchal history has been subject to until today—not only in western countries. With all these texts of women philosophers available, the methodical approach will change. It will become evident that it was not the "gendered" interest of women to rewrite the history of philosophy, but the universal philosophical necessity to criticize a history that was, in many regards, not dedicated to universality but defended particular interests. The male-streamed history of philosophy must be blamed for disguising its gendered interest behind its universal claims.

Re-reading the history of philosophy with this new attitude, we can approach and return to the genuine task of philosophy: putting one's own convictions at stake, testing and analyzing the reliability of content and method, critically observing oneself so as not to idolatrize or follow prejudices. Philosophy knows no country, no nation, no sex, it is the only discipline dedicated to retracing what can be known.

The philosophical challenge is to be critical of what we receive and what we are subjected to. Philosophy is true to itself as the foundation of all scientific knowledge only as long as it is able to be the foundation of critical thinking. Philosophy's task is to rethink its own narratives on truth and reliability. The history of women philosophers is *one* of several possible critical philosophical endeavors to *advance* the potential of philosophy. This critical capacity to put everything at stake, even itself, is what makes philosophy foundational to knowledge and science. If philosophy is not willing to serve the aim of providing this foundational capacity, to rethink, reshape and question its own basis, *it fails its genuine task*.

Great philosophical minds dare to question philosophical traditions. Great female and male philosophers have identified the unwillingness of many established philosophers to question their own beliefs. Women philosophers have pointed to decisive failures in the history of philosophy and its methods. The feminist éclat in the philosophy of the twentieth century is nothing more than the outburst of a long-suppressed voice that is now being heard.

Reading the history of women philosophers as a history in its own right and with its own topics and answers must be understood as a genuine part of the philosophical tradition as well as being a necessary instrument of critique and reflection for "institutionalized" history. Beyond women philosophers' unveiling of the particularized interest of male-gendered people in suppressing women, it is their uppermost concern to establish the realm for universally minded thoughts. Women have contributed to the history of philosophy in both ways: responding to the great philosophical questions of human history by writing on metaphysics, epistemology, logic, politics, ethics and more, and women have contributed to philosophy when pointing out their exclusions. Both ways added new contents, unheard of up till then, both in the name of philosophy and the search for truth, none of it exclusively confined to women's interests.

4.7 The History of Philosophy—Instrument of Critical Revision

The history of philosophy is not a bookshelf where the books are stacked on top of one another and the book on top contains all the underlying ones. Every historical interpretation includes philosophical reflections that are organized in intellectual systems of relationships that "glue" these imagined worlds together, aiming to make more coherent what we have grasped of the *real* world, which is, as we know and accept, ineffable. Each good philosophy offers its own explanation of how the facts in the world are "glued" together and each of these proposals has its strengths and weaknesses, its failures and its insights. We only know in part; we can never understand fully. However, this partial knowledge should not become the only narrative. How could philosophy ever move forward if it had no access to the real world or to the facts which force us to order and reorganize what we know, again and again, adding interpretations and increasing understanding, continually striving to improve? If the history of philosophy were only a collection of different narratives, adding women's history would be an additional narrative. But this is not the way of philosophy. To add the knowledge of women philosophers is to add, not a narrative, but a substantial part of our philosophical heritage to explain this world. Women philosophers have pointed out many of the "false glues" of wrong presumptions, showing that much is wrong which is still part of a history, so great and wonderful in its intention, and often weak and misused in its presentation.

The history of philosophy, as it is challenged by the inclusion of women philosopher's writings, is no longer the one we were used to. The oppression of women philosophers served to erect and to maintain a culturally sexualized and religiously supported male dominance, an ostracism that was due to violence and not to reasonable need. Re-reading and reevaluating history is thus also much more than a contextual reading of the relevant conditions of a specific time period. Doing history as a methodic approach to philosophy is a unique and indispensable means to widen the insights of the present. The history of philosophy is, however, no quarry of ideas where you select the material that provides the concepts you are interested in and give it the interpretation you have gained from your present ideas, thereby feeding prevalent prejudices even more. This narrow-minded approach is a methodic misuse of history. Much more than a quarry where you obtain the material you asked for, supporting and nourishing the opinion you have already formed, the reading and re-reading of history provides an instrument for critically analyzing and even refuting what may have already convinced you. It is one of the important means of cutting through the veils of ignorance of our present convictions and to overcome their limitations.

This path can be seen as a mine where you find yourself lost, where nothing seems to have the form you would have liked to find support. Everything this journey delivers may turn out differently than you expected. And the more foreign the ideas one is able to embrace, the more revealing their reception within the history of ideas can become. The mind widens as it develops the ability to embrace different points of view. Single facts do not allow a justified turn in the interpretational direction. The sense of history in philosophy is the analysis of the structure of complex coordinated thoughts and their connected conclusions. There is no richer way to experience philosophy as a *kind of reality* in its ideal sense than to approach it in its historical content.

The rise of women philosophers' participation in philosophy is connected to the re-reading of its history. This was the case in Renaissance philosophy. It also happened when women in the early modern era investigated the foundations of their own cultural backgrounds and when feminists started to rethink the history of philosophy. Women philosophers began to do research on the history of philosophy, although they had been told there was nothing of importance to be found written by women philosophers. The study of history revealed the contrary to be true. The history of philosophy opens up a rich world that unveils many standard claims as lies.

The study of the history of philosophy has always been an elemental part of understanding the ideas of philosophy. These philosophical ideas present a world of alternative explanations of philosophical problems and arguments. Reading history is not a closed circle where particular ideas come back as a useful updating service. The study of women philosophers opens up a realm of freedom where views alternative to commonly accepted ones can be pursued. Where else can we discover a universe of ideas, some of which we have possibly never heard of or never stepped into and would never have otherwise been able to encounter? Such a wealth of new ideas is offered through the study of the history of women philosophers.

References

Acker, C. (2013). *Femmes, fêtes et philosophie en Grèce ancienne*. Paris (L'Harmattan).

Alarabi, N. (2012). Constance Naden's philosophical poetry. *Literature Compass, 11*, 848–860.

Allen, P. (1997a). *The pluralists. Aspasia. The Concept of Woman. The Aristotelian Revolution, 750 B.C.–A.D. 1250*. Columbia, South Carolina: University of South Carolina Press.

Allen, P. (1997b). *The concept of women, vol. 2: The Early Humanist Reformation, 1250–1500*. Wm. B. Eerdmans Publishing Company.

Annas, J. (1976). Plato's Republic and Feminism. *Philosophy, 51*(197), 307–321.

Asimopulos, P. (2011). The authenticity of Pericles' funeral oration in the European historical thought. *Facta Universitatis Series Philosophy, Sociology, Psychology and History, issue, 2*, 233–239.

Astell, M. (1986). *The first English feminist. Reflections upon marriage and other writings*. B. Hill (Ed.), Aldershot: Maurice Temple Smith.

Astell, M. (2013). *The Christian religion, as professed by a daughter of the Church of England*. J. Broad (Ed.), Toronto: Centre for Reformation and Renaissance Studies and Iter.

Bar On, B.-A. (Ed.) (1994). *Engendering origins: Critical feminist readings in Plato and Aristotle*. Albany: State University of New York Press.

Beaney, M., & Chapman, S. (2017). "Susan Stebbing". In E. N. Zalta (Ed.), *The Stanford Encyclopedia of Philosophy* (Summer 2017 Edition). https://plato.stanford.edu/archives/sum2017/entries/stebbing. Accessed 28 May 2019.

Beauvoir, S. de (2011). *The second sex*. London: Random House.

Beiser, F. C. (2012). *Diotima's children. German aesthetic rationalism from Leibniz to Lessing*. Oxford: Oxford Universal Press.

Bello, A. A. (2002). *Hedwig Conrad-Martius and the phenomenology of nature*. Dordrecht: Springer.

Blair, E. (2012). *Plato's dialectic on woman. Equal, therefore inferior*. New York: Routledge.

Bodkin, R. G. (1999). "Women's agency in classical economic thought. In A. Smith, H. Taylor Mill, & J. S. Mill (Eds.), *Feminist Economics*, vol. 5, issue 1, pp. 45–61.

Bos, E.-J. (2010). Princess Elizabeth of Bohemia and Descartes' Letters (1650–1665). *Historia Mathematica, issue, 3*, 485–502.

Britting, J. (2004). *Ayn Rand*. New York: Overlook Duckworth.

Brucker, J. (1745). *Pinacoteca scriptorium nostra aetate literis illustrium*, vol. IV. Augsburg (J. J. Haidium).

Calcagno, A. (2007). *The philosophy of Edith Stein*. Pittsburgh: Duquesne University Press.

Calcagno, A. (Ed.) (2018). *Gerda Walther's phenomenology of sociality, psychology, and religion*. Dordrecht (Springer Women in the History of Philosophy and Science, no. 2).

Calkins, M. W. (2010). *The persistent problems of philosophy*. Norwood: Nabu Press.

Cavarero, A. (1995). *In spite of Plato. A feminist rewriting of ancient philosophy*. New York: Routledge.

Cavendish, M. (2001). *Observations upon experimental philosophy*. E. O'Neill (Ed.), Oxford: Cambridge University Press.

Cavendish, M. (2003). *Political writings*. S. James (Ed.), Cambridge: Cambridge University Press.

Cavendish, M. (2004). *The blazing world and other writings*. K. Lilley (Ed.), New York: Penguin Books.

Ceranski, B. (1996). *Und sie fürchtet sich vor niemandem: Die Physikerin Laura Bassi (1711–1778)*. Frankfurt; New York: Campus Verlag.

Chapman, S. (2013). *Susan Stebbing and the language of common sense*. Basingstoke: Palgrave Macmillan.

Ciempiel, A. (2008). Wise and devoted or shrewd and shameless? The true face of Aspasia of Miletus. *Scripta Classica, issue, 5*, 25–31.

Conley, J. J. (2002). *The suspicion of virtue: Women philosophers in neoclassical France*. New York, Ithaca: Cornell University Press.

Conway, A., (1992). *The Conway letters. The correspondence of Anne, viscountess Conway, Henry More, and their friends, 1642–1684*. S. Hutton & M. Hope Nicolson (Eds.), Oxford: Clarendon Press.

Corrigan, G., & Kevin (1989). *The life of saint Macrina*. Eugene, Or: Wipf & Stock.

Coventry, L. (1997). Philosophy and Rhetoric in the Menexenus. *The Journal of Hellenic Studies, issue, 109,* 1–15.

Craciun, A. (2002). *A Routledge literary sourcebook on Mary Wollstonecraft's A vindication of the rights of woman*. New York: Routledge.

Daigle, C., & Golomb, J. (Eds.) (2009). *Beauvoir and Sartre. The Riddle of Influence*. Indiana: Indiana University Press.

Dascăl, R. (2005). On Christine de Pizan's Feminism. *Gender Studies, issue, 4,* 104–118.

De Baar, M. et al. (Eds.) (1996). *Choosing the better part. Anna Maria van Schurman (1607–1678)*. Dordrecht; Boston; London: Kluwer Academic Publishers.

Deakin, M. (2007). *Hypatia of Alexandria*. Mathematician and martyr. Amherst Prometheus Books.

Du Châtelet, É. (2007). *Selected philosophical and scientific writings*. In J. P. Zinnser & I. Bour (Eds.), Chicago; London: University of Chicago Press.

Du Châtelet, É. (2011). *Examens de la Bible*. Bertram Eugene Schwarzbach (Ed.), Paris: Champion.

Du Châtelet, É. (2018). *La correspondence d'Emilie du Châtelet.* In U. Kölving & A. Brown (Eds.), Ferney (Voltaire Centre International d'Études du XIIIe Siècle).

Du Châtelet, É., & Newton, I. (2015). *Principes mathématiques de la philosophie naturelle*. La traduction française des Philosophiae naturalis principia mathematica. 2. vol., Édition critique du manuscrit par M. Toulmonde. Ferney (Voltaire Centre International d'Études du XIIIe Siècle).

Dzielska, M. (1995). *Hypatia of Alexandria*. Cambridge: Harvard University Press.

Ettinger, E. (1997). *Hannah Arendt and Martin Heidegger.* New Haven: Yale University Press.

Elisabeth of Bohemia (2007). *The correspondence between Princess Elizabeth of Bohemia and René Descartes*. Edited by Lisa Shapiro, Chicago: University of Chicago Press.

Falco, M. J. (Ed.) (1996). *Feminist interpretations of Mary Wollstonecraft*. University Park: Pennsylvania State University Press.

Farnsworth, B. (2010). Conversing with Stalin, surviving the terror: The diaries of Aleksandra Kollontai and the internal life of politics. *Slavic Review, 4,* 944–970.

Fonte, M. (1997). *The worth of women. Wherein is clearly revealed their nobility and their superiority to men*. V. Cox (Ed.), Chicago, Illinois: University of Chicago Press.

Frank, G. (2000). Macrina's scar. *Journal of Early Christian Studies, 4,* 511–530.

Frankel, L. (1989). Damaris Cudworth Masham. *Hypatia, 1,* 80–90.

Frankel, L. (1991). Damaris Cudworth Masham. In M. E. Waithe (Ed.), *A History of Women Philosophers*, vol. III. Dordrecht: Springer, pp. 73–85.

Furumoto, L. (1994). Christine Ladd-Franklin's color theory: Strategy for claiming scientific authority? *Annals New York Academy of Sciences, 727*(1), 91–100.

Gilman, C. P. (1975). *Women and economics: A study of the economic relation between men and women as a factor in social evolution.* New York: Gordon Press.

Goodman, K. (1999). *Amazons and apprentices: Women and the German Parnassus in the early enlightenment*. Rochester: Camden House Publishing.

Goodman, K. (2009). *Adieu Divine Comtesse: Luise Gottsched, Charlotte Sophie Gräfin Bentinck und Johann Christoph Gottsched in ihren Briefen*. Würzburg: Königshausen & Neumann.

Gouges, O. de (1979). The declaration of the rights of woman. In H. B. Applewhite, M. D. Johnson & D. G. Levy (Eds.), *Women in Revolutionary Paris, 1789–1795: Selected Documents Translated With Notes and Commentary*. Urbana: University of Illinois Press, pp. 87–96.

Gouges, O. de (2013). *Ainsi soit Olympe de Gouges: La déclaration des droits de la femme et autres textes politiques*. Benoîte Groult (Ed.), Paris (Grasset. Impr.).

Gournay, M. Le, J. de (2002). *Apology for the woman writing and other works*. R. Hillman & C. Quesnel (Eds.), Chicago: University of Chicago Press.

Grafton, A. (1997). From apotheosis to analysis: Some late Renaissance histories of classical astronomy. In D. R. Kelley (Ed.), *History and the Disciplines: The Reclassification of Knowledge* (pp. 261–276). Rochester New York: University of Rochester Press.

Green, K. (2014). *A history of women's political thought in Europe, 1700–1800*. Cambridge: Cambridge University Press.

Green, K., & Hagengruber, R. (Eds.) (2015). The history of women's ideas. *The Monist* 98 (1).

Gregory, B. N. (1989). *The life of saint Macrina*. K. Corrigan, transl., Eugene, Oregon: Wipf & Stock.

Hagengruber, R. (Ed.) (2011). *Émilie du Châtelet between Leibniz and Newton*. New York: Springer.

Hagengruber, R. (2015). Soziale Organisation und enzyklopädisches Interesse. Campanellas metaphysische Grundlegung der Sonnenstadt, In O. Höffe (Ed.) *Politische Utopien der Neuzeit* (pp. 139–157). Berlin: De Gruyter.

Hagengruber, R. (2018). Relocating women in the history of philosophy and science: Emilie Du Châtelet (1706–1749), Laura Bassi (1711–1778), and Luise Gottsched (1713–1762) in Brucker's *Pinacotheca*. In S. Plastina, E. M. De Tommaso (Eds.), Filosofe e scienziate in età moderna.*Bruniana&Campanelliana*, Supplemento.

Hagengruber, R., & Rodrigues, A. (Eds.) (2010). *Von Diana zu Minerva. Philosophierende Aristokratinnen des 17. und 18. Jahrhunderts*, Berlin: Akademie Verlag.

Hagengruber, R., & Hecht, H. (Eds.) (2019). *Émilie Du Châtelet und die deutsche Aufklärung*. Wiesbaden: Springer.

Harding, S. (1991). *Whose science? Whose knowledge? Thinking from women's lives*. New York: Cornell University Press.

Harth, E. (1992). *Cartesian women: Versions and subversions of rational discourse in the old regime*. New York, Ithaca: Cornell University Press.

Haskins, E. (2005). Philosophy, rhetoric, and cultural memory. Rereading Plato's Menexenus and Isocrates' Panegyricus. *Rhetoric Society Quarterly, 1*, 25–45.

Henry, M. M. (1995). *Prisoner of history. Aspasia of Miletus and her biographical tradition*. New York; Oxford: Oxford University Press.

Hildegard, B. (1997). *The book of the rewards of life: Liber vitae meritorum* (B. W. Hozeski transl.). New York: Garland.

Hutton, S. (1993). Damaris Cudworth, lady Masham. Between Platonism and enlightenment. *British Journal for the History of Philosophy, 1*, 29–54.

Hutton, S. (2004). *Anne Conway. A woman philosopher*. Cambridge: Cambridge University Press.

Hutton, S. (2012). Women philosophers and the early reception of Descartes. Anne Conway and Princess Elizabeth. In T. M. Schmaltz (Ed.), *Receptions of Descartes. Cartesianism and Anti-Cartesianism in Early Modern Europe*. New York: Routledge, pp. 3–21.

Ilsley, M. H. (1963). *A daughter of the Renaissance. Marie le Jars de Gournay, her life and works*. The Hague: Mouton.

Irigaray, L. (1992). Sorcerer Love. A Reading of Plato's Symposium, Diotima's Speech. In S. L. Bartky & N. Fraser (Eds.), *Revaluing French Feminism. Critical Essays on Difference Agency, and Culture*. Bloomington: Indiana University Press, 64-76.

Karafili Steiner, E. (2014). *Called to civil existence. Mary Wollstonecraft's A vindication of the rights of woman*. New York.

Kerby-Fulton, K. (2010). Hildegard of Bingen. In *Medieval Holy Women in the Christian Tradition c.1100–c.1500*. Alastair Minnis and Rosalynn Voaden (Eds.), Turnhout (pp. 343–369), Belgium: Brepols.

King, M. L. (1991). *Women of the renaissance*. Chicago: University of Chicago Press.

King, M. L. (2005). *Humanism, Venice, and women. Essays on the Italian Renaissance*. Aldershot & Burlington: Ashgate.

Klens, U. (1996). *Mathematikerinnen im 18. Jahrhundert: Maria Gaetana, Agnesi, Gabrielle-Emilie du Châtelet, Sophie Germain. Fallstudien zur Wechselwirkung von Wissenschaft und Philosophie im Zeitalter der Aufklärung*. Pfaffenweiler: Centaurus.

Kölving, U., & Courcelle, O. (2008). *Émilie Du Châtelet: Éclairages & documents nouveaux*. Ferney (Voltaire Centre International d'Études du XIIIe Siècle).

Langdon, K. (2002). *The political theory of Christine de Pizan*. Aldershot: Ashgate.

Lazaroiu, G. (2013). Pythagorean women. *Journal of Research in Gender Studies*, 2, 147.

Lebech, M. (2015). *The philosophy of Edith Stein. From Phenomenology to Metaphysics*: Peter Lang Verlag.

Lerius, J. (2017). Hildegard von Bingen on autonomy. In S. Berges, A. L. Siani (Eds.), *Women philosophers on autonomy: Historical and contemporary perspectives,* (pp .17-31), London: Routledge, pp. 17-31.

Lerner, G. (1993/2013). From the creation of feminist consciousness: From the middle ages to eighteen-seventy. In S. M. Harris & L. K. Hughes (Eds.), *A Feminist Reader. Feminist thought from Sappho to Satrapi*, vol. IV (pp. 302-337) Cambridge: Cambridge University Press.

Lloyd, G. (1984). *Man of reason*. London: Methuen Publishing.

Lokaneeta, J. (2001). Alexandra Kollontai and Marxist veminism. *Economic and Political Weekly, 17,* 1405–1412.

Luft, S., & Hagengruber, R. (2018). *Women phenomenologists on social ontology. We-experiences, communal life and joint action*. Dordrecht (Springer History of Women Philosophers and Scientists no. 1).

Madigan, S., & O'Hara, R. (1992). Short-term memory at the turn of the century: Mary Whiton Calkins' memory research. *American Psychologist, 2,* 170–174.

Maier-Katkin, D. (2010). *Stranger from abroad. Hannah Arendt, Martin Heidegger, friendship and forgiveness*. New York (W.W. Norton & Company).

Marshall, R. K. (1998). *The winter queen. The Life of Elizabeth of Bohemia, 1596–1662*. Edinburgh: National Galleries of Scotland.

Masham, D. C. (2004). *The philosophical works*. introduced by J. G. Buickerood. Bristol: Thoemmes Continuum.

Mazzotti, M. (2007). *The world of maria Gaetana Agnesi, mathematician of God*. Baltimore, MD: Johns Hopkins University Press.

McAlister, L. L. (1989). Some remarks on exploring the history of women in philosophy. *Hypatia, 4*(1–5), 3.

McAlister, L. L. (1992). Dr. Gerda Walther: German feminist philosopher. In M. Pellikaan-Engel (Ed.), *Against patriarchal thinking* (pp. 65–69). Amsterdam: VU University Press.

McDonald, D. N. (2007). Differing conceptions of personhood within the psychology and philosophy of Mary Whiton Calkins. In *Transactions of the Charles S. Peirce Society: A Quarterly Journal in American Philosophy* 4, pp. 753–768.

McWeeny, J. (2011). Princess Elizabeth and the mind–body problem. In S. Barbone & M. Bruce (Eds.), *Just the Arguments. 100 of the most important arguments in Western philosophy* (pp. 297–300). Malden, Mass: Wiley-Blackwell.

Ménage, G. (1690/1984). *The history of women philosophers*. B. H. Zedler (Ed.), Lanham, MD: University Press of America Lanham.

Merchant, C. (1979). The vitalism of Anne Conway. Its impact on Leibniz's concept of the monad. *Journal of the History of Philosophy, 3,* 255–269.

Mill, H. T. (1998). T*he complete works of Harriet Taylor Mill*. Ed. by J. E. Jacobs. Bloomington: Indiana University Press.

Mill, H. T., & Mill, J. S. (1970). *Essays on sex equality*. A. S. Rossi (Ed.), Chicago: University of Chicago Press.

Muller, C., & Bocquet, J.-L. (2012). *Olympe de Gouges*. Paris: Casterman.

Munslow, A. (2012). *A history of history*. London: Routledge.

Naden, C. C. (1890). *Induction and deduction: A historical and critical sketch of successive philosophical conceptions respecting the relations between inductive and deductive thought and other essays*. London: Bickers.

Nails, D. (2002). *The people of Plato*. Indianapolis: A Prosopography of Plato and Other Socratics.

Neumann, H. (1965). Diotima's Concept of Love. *The American Journal of Philosophy 86*, 33–59.

Nogarola, I. (2004). *Complete writings. Letterbook, Dialogue on Adam and Eve, Orations*. M. King & D. M. Robin (Eds.). Chicago: University of Chicago Press.
Nye, A. (1989). The hidden host: Irigaray and Diotima at Plato's 'Symposium'. *Hypatia, 3,* 45–61.
Olympe, G. de (1979). The declaration of the rights of woman. In H. B. Applewhite, M. D. Johnson & D. G. Levy (Eds.), *Women in Revolutionary Paris, 1789–1795: Selected Documents Translated With Notes and Commentary* (pp. 87–96). Urbana: University of Illinois Press.
Olympe, G. de (2013). *Ainsi soit Olympe de Gouges: La Déclaration des droits de la femme et autres textes politiques*. B. Groult (Ed.), Paris: Grasset.
O'Mahony, P. (2010). The origin of the olive. On the Dynamics of Plato's 'Menexenus'. *Polis: The Journal of Ancient Greek Political Thought*, 1, 38–57.
Ornes, S. (2009). *Sophie Germain*. Greensboro, NC: Morgan Reynolds Publishing.
Peirce, C. S., & Welby, V. (2001). *Semiotic and significs. The correspondence between Charles S. Peirce and Victoria Lady Welby*. C. S. Hardwick (Ed.), Elsah, IL: Arisbe Associates.
Perry, R. (1986). The celebrated Mary Astell: An early English feminist. Chicago: University of Chicago Press.
Petrilli, S. (2009). Signifying and understanding: Reading the works of Victoria Welby and the signific movement. Berlin: de Gruyter.
Pizan, C. de (1999). *The book of the city of ladies*. R. Brown-Grant (Ed.), London: Penguin Books.
Pomeroy, S. B. (1984). *Women in classical antiquity*. New York: Schocken Books.
Pomeroy, S. B. (2002). *Spartan women*. Oxford: New York: Oxford University Press.
Pomeroy, S. B. (2013). *Pythagorean women. Their history and writings*. Baltimore John Hopkins University Press.
Porter, C. (2014). *Alexandra Kollontai: A biography*. Chicago: Haymarket Books.
Prasch, R. E., & Sheth, F. A. (1996). Charlotte Perkins Gilman. Reassessing her significance for feminism and social economics. *Review of Social Economy*, 3, 323–335.
Rand, A. (1961). *For the new intellectual*. The Philosophy of Ayn Rand. New York (Random House).
Rauschenbach, B. (2000). *Der Traum und sein Schatten. Frühfeministin und geistige Verbündete Montaignes: Marie de Gournay und ihre Zeit*. Königstein/Taunus: Ulrike Helmer Verlag.
Ready, K. J. (2002). Damaris Cudworth Masham, Catharine Trotter Cockburn, and the feminist legacy of Locke's theory of personal identity. *Eighteenth-century Studies*, 4, 563–576.
Rees, E. L. E. (2003). *Margaret Cavendish. Gender, genre, exile*. Manchester: Manchester University Press.
Reeser, T. W. (2016). *Setting Plato straight: Translating ancient sexuality in the renaissance*. Chicago: University of Chicago Press.
Reisel Gladstein, M., & Sciabarra, C. M. (Eds.) (2003). *Feminist interpretations of Ayn Rand*. University Park, Pennsylvania: Pennsylvania State University Press.
Rogers, D. G. (2005). *America's first women philosophers. Transplanting Hegel, 1860–1925*. London: Bloomsbury Publishing.
Salkever, S. G. (1993). Socrates' Aspasian oration. The play of philosophy and politics in Plato's Menexenus. *The American Political Science Review, 1,* 133–143.
Schmitt, C., Skinner, Q., & Kessler, E. (Eds.). (1988). T*he Cambridge history of renaissance philosophy*. Cambridge: Cambridge University Press.
Schott, R. M. (2007). Feminism and the history of philosophy. Reprinted In L. Alcoff & E. F. Kittay (Eds.) *The Blackwell Guide to Feminist Philosophy* (pp. 43–59). Hoboken: Wiley-Blackwell.
Schröder, H. (1995). *Olympe de Gouges. Mensch und Bürgerin. Die Rechte der Frau*. Aachen: Ein-Fach-Verlag.
Shapiro, L. (1999). Princess Elizabeth and Descartes. The union of soul and body and the practice of philosophy. *British Journal for the History of Philosophy, 3,* 503–520.
Sheridan, P. (2007). Reflection, nature, and moral law. The extent of Catharine Cockburn's Lockeanism in her defence of Mr. Locke's Essay. *Hypatia, 3,* 133–151.
Sherman, C. L. (2013). *Reading Olympe de Gouges*. New York: Palgrave Macmillan.

Smith, A. J. (1992). *Reading Julia Kristeva: Estrangement and the female intellectual.* A thesis submitted in the fulfilment of the requirements of the Doctorate in Philosophy in the University of Canterbury.
Smith, H. L. (2007). Women intellectuals and intellectual history: Their paradigmatic separation. *Women's History Review, 16,* 353–368.
Springborg, P. (1995). Mary Astell (1666–1731), Critic of Locke. *American Political Science Review, 89*(3), 621–633.
Stegmann, H. (1910). Aspasia. In G. Wissowa & W. Kroll (Eds.), *Paulys Realencyclopädie der Classischen Altertumswissenschaft,* vol. 3 (pp. 1716–1722). Stuttgart: Metzler.
Stein, E. (1989). *On the problem of empathy.* L. Gelber (Ed.), Washington, DC: ICS Publications.
Stein, E. (1990–2014). *Edith-Stein-Gesamtausgabe,* vol. 1–10. Mariéle-Claudia Wulf, Hanna-Barbara Gerl-Falkovitz et al. (eds.), Freiburg im Breisgau (Verlag Herder).
Stein, E. (2006). *An investigation concerning the state.* L. Gelber (Ed.), Washington, DC: ICS Publications.
Stopczyk, A. (1980): *Was Philosophen über Frauen denken.* München: Matthes & Seitz Verlag
Strohmeyr, A. (2006). *Sophie von La Roche.* Leipzig (Reclam): Eine Biografie.
Szapuová, M. (2006). Women in philosophy. The case of Harriet Taylor Mill. *Human Affairs, 2,* 133–143.
Theano. (2010). *Briefe einer antiken Philosophin.* Kai Brodersen (Ed.), Stuttgart (Reclam).
Todd, J. (2000). *Mary Wollstonecraft: A revolutionary life.* New York: Columbia University Press.
Tuana, N. (Ed.) (1994). *Feminist interpretations of Plato.* University Park, Pennsylvania: The Pennsylvania State University Press.
Villa, D. (1995). *Arendt and Heidegger.* The fate of the political. Princeton: Princeton University Press.
Waithe, M. E. (Ed.) (1987–1995). *A history of women philosophers,* vol. I–IV. Dordrecht: Kluwer.
Weiss, P. A. (2004). Mary Astell: Including Women's voices in political theory. *Hypatia, 3,* 63–84.
Weissberg, L. (2000). *Rahel Varnhagen: The life of a Jewess.* Baltimore: The Johns Hopkins University Press.
Wieland, C. M. (1796). Die pythagoreischen Frauen. Ehrenrettung der Aspasia, Julia und Faustina. In *Sämtliche Werke,* vol. 24. Leipzig: Göschen.
Whitaker, K. (2003). *Mad Madge. The extraordinary life of Margaret Cavendish, duchess of Newcastle, the first woman to live by her pen.* New York: Basic Books.
Wider, K. (1986). Women philosophers in the ancient Greek world. Donning the mantle. *Hypatia, 1,* 21–62.
Willard, C. C. (1990). *Christine de Pizan.* New York (Persea): Her Life and Works.
Witt, C., & Shapiro, L. (2019). Feminist history of philosophy. In E. N. Zalta (Ed.), The Stanford Encyclopedia of Philosophy (Fall 2018 Edition), https://plato.stanford.edu/archives/fall2018/entries/feminism-femhist Accessed 28 May 2019.
Wollstonecraft, M. (1989). *The works of Mary Wollstonecraft.* M. Butler, E. Rees-Mogg & J. Todd (Eds.), London: Pickering.
Yates, F. A. (1983). *The Italian academies. In Renaissance and reform: The Italian contribution.* Collected essays. London: Routledge, pp. 6–29.
Zinnser, J. & Candler Hayes, J. (Eds.) (2006). *Emilie Du Châtelet: Rewriting enlightenment philosophy and science.* Oxford: Voltaire Foundation.

Part II
Rewriting the History

Chapter 5
The Goddess and Diotima: Their Role in Parmenides' Poem and Plato's Symposium

Vigdis Songe-Møller

Abstract While female characters play a central part in Greek comedies and tragedies, this is not the case in Greek philosophical texts. There are, however, two important exceptions: in Parmenides' poem and Plato's *Symposium* female characters—an unnamed goddess and the priestess Diotima—have unique access to philosophical truth, which they convey to their male pupils. This chapter poses the following question: Why did Plato and Parmenides choose female characters as a precondition for the philosophical quest for knowledge? It is argued that both the goddess and Diotima tell a truth that is beyond the reach of ordinary human knowledge, from a perspective of the Other. While Parmenides uses a female non-human character to expel everything female from true thinking, Plato uses Diotima to destabilize the notions of male and female and thus to point towards a non-gendered subject of philosophy.

5.1 Introduction

Among the texts that have survived from Greek antiquity, there are few written by women. Sappho is a great exception. In addition to her poems, we have little else than some fragments by Pythagorean women. It is hardly an exaggeration to say that the ancient Greek texts that have been handed down to us are written by men, and for men, within a dominantly male society, where women were not only subordinated to men—be it to their fathers, husbands, or other male relatives—but also bereft of political rights. The Greek *polis* has accordingly been called a "men's club" where women were excluded and thus to a certain extent made invisible (Vidal-Naquet 1986, p. 206). Both the subordination and exclusion of women are explicitly thematized in several of these texts, for instance in the political and biological writings of Aristotle: women's political and biological subordination is not only taken for granted but also philosophically and quasi empirically argued for, to the extent that

V. Songe-Møller (✉)
Department of Philosophy, University of Bergen, Bergen, Norway
e-mail: Vigdis.Songe-Moller@uib.no

S. Thorgeirsdottir and R. E. Hagengruber (eds.), *Methodological Reflections on Women's Contribution and Influence in the History of Philosophy*, Women in the History of Philosophy and Sciences 3, https://doi.org/10.1007/978-3-030-44421-1_5

woman's contribution in the reproduction process is reduced to a minimum.[1] As has been shown in feminist scholarship over the last decades, also in texts where the relation between men and women is not as explicitly discussed as by Aristotle, a subordination and/or exclusion of women form part of the underlying argument: a "fear of diversity", or a "dream of the superfluity of women", may be said to run through much of the ancient Greek literature.[2]

While both in Greek comedies and tragedies, female characters play an important part, this is not the case in philosophical literature. There are, however, two important exceptions. In Parmenides' poem, a goddess plays the central part: she is the one who speaks the truth to the young man who visits her. In the *Symposium*, Plato gives Diotima an equivalent role: she is the one who knows the truth about love and teaches this to the young, erotic philosopher. While these two female characters seem to be the precondition for the philosophical quest for truth, they themselves are, as Kristin Sampson (2013, p. 105) has pointed out, placed in an outside position of this quest.[3]

Before I start my discussion of the role of the Parmenidean goddess and Diotima I have to mention the third exception in the use of female characters in the ancient philosophical literature. In addition to Diotima, Plato gives voice to one other woman: in the *Menexenus*, Socrates narrates a funeral oration which Aspasia had delivered the day before, and Socrates even calls Aspasia his teacher, obviously in the art of rhetoric (*Menexenus* 236c). There is thus an apparent parallel between Aspasia and Diotima, whom Socrates calls his teacher in matters of love (*Symposium*, 201d). While Diotima's existence is uncertain and may be an invention of Plato, Aspasia was a well-known woman in Athens. She was the companion of Athens' most powerful man, Pericles, for about twenty years till his death in 429, and she was frequently portrayed in contemporary literature. In comic drama, she was made into a sexual figure, a whore and a brothel keeper.[4] In the *Menexenus* she has a different role, but also Plato seems to, at least indirectly, make use of her reputation as a seductive woman and her familiarity with men. Aspasia's funeral oration is a highly seductive one, flattering the Athenians, both dead and living. Plato also makes a point of the familiarity between Socrates and Aspasia: Socrates tells that he "narrowly escaped a

---

[1] See for instance *Generation of Animals*, book 2, 728a 18–20: 'A woman is as it were an infertile male; the female, in fact, is female on account of a certain inability, being incapable of concocting the nutriment in its last stage into semen'.

[2] Cf. these titles: *Fear of Diversity. The Birth of Political Science and Greek Thought* (Saxonhouse 1995); The Dream of a World without Women: Poetics and the Circles of Order in the *Theogony Prooemium* (Arthur 1983); The Greek Dream of a Womanless World (Songe-Møller 2002, Chap. 1); and Jean-Pierre Vernant's phrase 'this dream of a purely paternal heredity never ceased to haunt the Greek imagination' (Vernant 1983, p. 134).

[3] Cf. also Sampson's remark: '… they [the goddess in Parmenides' poem and Diotima] both possess wisdom and tutor men in search for it' (Sampson 2013, p. 105). Also Saxonhouse (1995, p. 174, n. 16) mentions a possible relationship between Diotima and the goddess in Parmenides' poem, as does Cavarero (1995, p. 94).

[4] In Aristophanes *Acharnians*, there is a reference to 'two of Aspasia's whores' (527). For an analysis of the treatment of Aspasia in Attic comedy, see Henry (1995, 19–28).

beating" from her each time he failed to memorize her speech correctly (*Menexenus* 236b-c), and he seems to have been in the habit of visiting her (*Menexenus* 249d).[5]

There can be no doubt that Aspasia's speech in the *Menexenus* serves as an ironic critique, even a parody, of the Athenian funeral oration, especially Pericles' famous one, and its praise of the Athenian *polis*.[6] Thus, also the presentation of Aspasia as an authority and as Socrates' teacher in rhetoric must be understood as ironical. If there is a parallel between Aspasia and Diotima, it is one of inversion. Whereas Ditoma is a priestess and explicitly called a foreigner (*ksenê*, *Symposium* 201e) and thus stands apart from the erotic assembly of Athenian men in the *Symposium*, Aspasia, even though she is a woman from Miletus, is presented as a more or less integrated part of the Athenian intellectual world. Aspasia was, as Debra Nails puts it, "the de facto wife of Pericles" (Nails 2002, p. 59) and thus a family woman and also a mother, like other Athenian wives. As a priestess from Mantinea, Diotima is situated far away from this traditional Athenian way of living. Last but not least: while Socrates calls Diotima "wise" (*sophê*; *Symposium*, 201d) and her authority as a teacher in erotics is not questioned, Aspasia's narrated funeral oration is indirectly being mocked, and Socrates' praise of her as a teacher in rhetoric is thus undermined.

In the following, I will discuss in which way the relationship between Diotima and Socrates may be said to be parallel to that between the goddess and the young man who visits her. My main question will be: Why did Plato and Parmenides choose female characters as these two men's teachers, who both of them occupy an outside position to the men's quest for knowledge? In Parmenides' poem, the goddess is only one—although the central one—among several female divinities. A substantial part of my discussion of the poem will be devoted to these other divinities, which is essential for an understanding of Parmenides' extensive use of female figures in his endeavor to communicate truth.

5.2 The Role of the Goddess and Other Female Characters in Parmenides' Poem

When Parmenides wrote his poem, probably in the first half of the fifth century BC, the word "philosophy" did not exist, and it is therefore somewhat anachronistic to call him a philosopher and his poem a work of philosophy. Nevertheless, he started a way of thinking that not only influenced but also to a certain extent determined the intellectual endeavor that Plato worked out as a separate genre and called "philosophy". While before Parmenides, thinkers who were later included in the philosophical canon—such as Thales, Anaximander, and Heraclitus—asked questions about the origin of all things and tried to come to terms with the changing world, Parmenides simply

[5]Cf. Halperin (1990, 123). Halperin refers to other contemporary texts where the relation between Socrates and Aspasia allude to sexual matters, as Xenophon's *Memorabilia* (2.6.36), and *Oeconomics* (3.14).

[6]See especially Loraux (2006).

denied the notions of origin and change. According to Parmenides, there is neither an origin nor a past nor a future, but only a presence, a timeless "now". Or, to be precise: this refusal of a changing world is put in the mouth of a goddess. I believe it is no coincidence that Parmenides, who opens for a way of thinking that transcends the human world, needs a female character to speak the truth he is seeking.

In Parmenides' poem, there is one male character—the poem's only human being—but a lot of female ones. The male character, a young man (*kouros*), is the speaking subject of the first 23 lines of the poem, where he relates his extraordinary journey to the goddess. He begins his tale in this way:

> The mares that carry me as far as ever my spirit (*thymos*) may aspire,
> were escorting me, when they brought me and proceeded along the renowned road
> of the goddess (*daimôn*),[7] which brings a knowing mortal to all cities.
> On this road I was being brought, on it wise mares were bringing me,
> straining the chariot, and maidens (*kourai*) were guiding the way (28B1, 1–5).[8]

Already here, in the very first verses of the poem, Parmenides strongly emphasizes the female environment of the young aspiring man, who himself is in a passive situation, totally dependent on his female helpers—mares that carry him and maidens who guide him—in order to travel on the road which leads him to the goddess. The guiding maidens, who accompany him all the way to the goddess, turn out to be "daughters of the Sun", coming from "the house of Night" and moving towards light, "having pushed back the veils from their heads with their hands" (28B1, 10). These metaphors of light and unveiling indicate that the young man's journey is what we today would call a philosophical one, a philosophical quest for knowledge, or for truth. While the young man, although physically passive, is the one aspiring for knowledge and enlightenment, the mares and maidens—and all the other female figures in the poem—do not seem to have any such desire. But they are the preconditions for his philosophical quest.

Before the chariot reaches the house of the goddess, it arrives at a crucial location: "the gates of the roads of Night and Day" (28B1, 11), which *Dike* guards. As a goddess of justice, *Dike* watches over different kinds of boundaries, and in Parmenides' poem, she guards a decisive boundary between the world of humankind on the one side and the residence of the goddess on the other. There is obviously something extraordinary in the fact that the young man is permitted to cross this line, and the crossing itself is described as something frightful: after the maidens "with soft words" have persuaded *Dike* to open the gates, a "gaping chasm" (*xasma*) appears, through which the maidens steer the chariot. This "gaping chasm" alludes to Hesiod's descriptions both of Chaos

---

[7]This female *daimôn* is not the goddess (*thea*) who teaches the young man the truth about being. This *daimôn*, however, we will meet later, in the goddess' description of the false human opinions. See below.

[8]References to the fragments of Parmenides are given according to the numbering established by Diels and Kranz (1951). The translations are based on the following: Curd (2011), Coxon (2009), Tarán (1965).

(*xaos* meaning "chasm", "gap", "gulf") and of Tartaros, a fearful, bottomless abyss, "awful (*deinos*) even to the deathless gods" (*Theogony* 743).[9]

In Hesiod's *Theogony*, Chaos is, alongside Earth (*Gaia*), the first being (*Theogony*, 116–117). While Earth is a primal mother, analogous to a woman, Chaos, as the original cleft or chasm, is more like a primal feminine sexual symbol. It is therefore not far fetched when Andrea Nye interprets the "gaping chasm" in Parmenides' poem as a "gaping vaginal cleft" and *Dike*'s gates as "great double vaginal doors" (Nye 1990, p. 18). In his passage through the doors and the chasm, Parmenides has, according to Nye, "crossed into another world" and "is reborn" (Nye 1990, p. 19). The theme of birth is an important one, and I will come back to this. My main point here is that the gates and the gaping chasm separate the human and the divine world. The narration of the journey stops with the words: "There, straight through them [*Dike*'s gates], then, the maidens held the chariot and horses on the broad road" (28B1, 20–21). In the next verse, the young man is greeted by the goddess.

Why is the young man dependent on exclusively female, non-human characters to transcend the human world, which in the poem is characterized by its cities? It seems obvious that he could not have taken this journey by himself, or with the help of other human beings alone. He needed someone with non-human powers, some other than a human being. But why did these others have to be female? By using only female, non-human characters—the mares, the daughters of the Sun, and *Dike*—Parmenides is able to emphasize the distinct character of the traveler: a man, belonging to the male world of Greek cities. Taken together the female helpers represent a double Other of the male, human traveler: female and non-human. By this double otherness, Parmenides underscores that the narrated quest for knowledge belongs to man alone.[10]

A spectacular journey is left behind. The young man has described it in vivid terms, both visual and auditory, for instance, details of the chariot—how its axle made "the bright sound of a musical pipe" (28B1, 6)—and details of the huge, bronze gates, with "a lintel and a stone threshold" (28B1, 12). When the chariot has crossed the gaping chasm, there are no more descriptions. Or to put it in other words: there is nothing to describe. The motions, sounds, and visual, material things of the world he has left are no more. The mares and the maidens have disappeared. From this point onwards, the young man is in another, divine universe, where there is only himself and the goddess. Her words, or rather the young man's rendering of her words, occupy the rest of the poem. The young man only listens and does as the goddess commands him to do, namely to preserve and pass on her story (28B2, 1).

[9]The numbering refers to verse lines. For a translation of the *Theogony*, see Hesiod (1966).

[10]Several feminist scholars have commented on this striking aspect of the poem that the young man is the only male figure, assisted by several female figures, and also on the young man's passivity. See Saxonhouse (1995, pp. 39–40). Canters and Jantzen (2005, p. 12) comments that '[f]rom a feminist perspective it is ironic that females assist Parmenides in his quest for knowledge and yet are excluded from it themselves: having given him their aid, they are then dropped from consideration.' Cavarero (1995, p. 38) asserts that 'it is astounding that female figures inaugurate the route towards abstract thought, where philosophy celebrates its patriarchal glory'.

The goddess is nameless and cannot be identified with any divinity in earlier mythology. By contrast, the other divinities which occur in the poem—the daughters of the Sun, *Dike*, *Moira*, and *Ananke*—belong to the traditional world of Greek mythology. We, or Parmenides' readers, know who they are. The goddess, however, does not belong to any "world". She is nothing but her words. She is a speaking subject; there is no background, or context, for her words. The goddess speaks from nowhere, or perhaps rather: from an unknown—and unknowable—transcendent region. Why does Parmenides need to communicate his thoughts about being from this perspective? Why cannot he tell us his thoughts himself, from a human, male perspective? Before I try to answer these questions, I will briefly go into the goddess' teaching.[11]

> The goddess declares that the young man shall 'learn all things',
> both the unshaken heart of well-rounded truth (*alêtheia*)
> and the beliefs (*doxai*) of mortals in which there is no true trust (*pistis alêthes*).
> (28B1, 29–30)
> For the goddess, thinking and being are mutually dependent and mutually constitutive,
> for thinking and being are the same. (28B3)

One consequence of this claim is that where there is nothing, there is also nothing to be thought. The goddess thus tells the young man that

> It is necessary to say and to think what is, for there is being,
> but nothing is not. (28B6, 1–2)

I interpret this sentence as a kind of command to the young man. This is his mission, his mission as a thinker, or, as we might say, as a philosopher: to think and to put into words that which is.[12] On the first sight of it, this kind of task is close to what philosophers or poets try to live up to, but as the goddess develops her thoughts, we soon discover that what she calls "being", or "what is", is radically different from "what is", seen from a human perspective. We should note that the goddess does not put herself into words: being the anonymous subject of her words, she is outside of the kind of being she is talking about.

The basis for the goddess' argument is a radical distinction between being and not-being: "Is or is not" (28B8, 16). In fragment 8, which is the longest preserved part of the poem, the goddess describes the consequences of—and maybe the motivation for—this radical distinction. It starts like this:

> Just one story of a route
> is still left: that it is. On this route there are very many signs (*sêmata*):
> that being is ungenerated and imperishable,
> a whole and of one kind, unshaken and complete.
> Nor was it ever, nor will it be, since it is now, altogether one,

[11] For a detailed argument for my interpretation of Parmenides' poem, see Songe-Møller (2002, Chaps. 2 and 3).

[12] Here, I follow Heidegger (1971, part II) in his analysis of this fragment.

indivisible. For what birth will you seek out for it? (28B8, 1–6)

The first two signs of being, "ungenerated" (*ageneton*) and "imperishable" (*anôlethron*), may be translated as "unborn" and "immortal": as unborn being is also immortal. The last question, which obviously is a rhetorical one, indicates that birth is a problem of fundamental importance for the goddess. She confirms this assumption in the following verses, where she, over and over again, shows how absurd it would be to say and to think that there is birth since the notion of birth—birth from nothing, birth from something that is, birth in the past, and birth in the future—involves not-being. Where there is no birth, there is no death, and the goddess concludes:

Thus birth (*genesis*) has been extinguished and death is unheard of (28B8, 21).

The timeless, immortal being is thus secured through thinking, by keeping being pure, with no stain of not-being. Still, being seems to be a somewhat unruly entity, for the goddess is in need of several other goddesses in order to keep it safe from not-being: *Dike*, *Moira*, and *Ananke*. Like *Dike*, also *Moira* and *Ananke* are guardians of borders: in Greek mythology *Moira* guards the borders of life and death by presiding at childbirth (cf. *Symosium* 206d) and cutting the thread of life, and *Ananke*, meaning "constraint" or "necessity", is also a goddess of fate, keeping men forcefully in their place.[13] Their function in the goddess' construction of truth is to keep strong, unbreakable chains, or fetters, around being:

Therefore *Dike* did not loosen her shackles,
neither for coming to be nor for perishing,
but holds it [being] fast. (28B8, 13–15)
...For powerful *Ananke*
holds it [being] in the bondage of a limit which encircles being. (28B8, 30–31)
...For nothing else either is or will be
except being, since *Moira* fettered it to be whole and changeless. (28B8, 36–37)

In describing being as something that needs to be kept in place by harsh and unbreakable chains, the goddess seems to imply that being possesses an inner potential, or force, that seeks to overcome its limits. Being is thus not a static, or lifeless, entity, as often has been claimed. Being, which the goddess expresses by the verb "is" (*estin*), is not a thing, but must be thought of in its verbal sense, as the constitutive activity, or event, of being. But it is a kind of activity that never—thanks not only to divine thinking but also to the divine fetters around it—results in anything but pure, timeless being: not birth, not change, not death. Being just *is*, now, one and whole.[14] Why are the fetters around being held by female, and not male, divinities? My answer is simple and perhaps self-evident: what is kept out from being, is that for which woman, according to traditional Greek thinking, in literature both prior to and after

[13]In Plato's *Republic*, book X, 617b-c, the three *Moirai* (Lachesis, Clotho, and Atropos) are said to be daughters of *Ananke*, and in the *Symposium*, 197b, Agathon claims that *Ananke* was in power before *Eros* was born.

[14]For a more detailed analysis of being as a constitutive 'event' of being, see Songe-Møller (2002, pp. 43–47), where Parmenidean being is interpreted as a masculine (viril) activity.

Parmenides, is responsible. In *Theogony*, Hesiod calls Pandora, the first woman, the "death bringing sex" (591). According to Hesiod, women brought difference, sexuality and birth, and—as a consequence of birth—death into the world.[15] Female goddesses—and not male gods—seem to be best fitted to keep the goddess' first sign of not-being, namely birth, away from the pure, timeless, immortal being.

In her welcome to the young man, the goddess promised that he should learn also "the beliefs (*doxai*) of mortals in which there is no true trust" (28B1, 29–30). In these beliefs, characterized by difference, plurality and change, *eros* and birth play a central role. In accordance with the false human beliefs, the ruler of the world is an unnamed female divinity, a *daimôn*, who is situated in the middle of the universe. This female *daimôn* "devised *Eros* as the very first of all gods" (28B13), and she is also responsible for the "abominable birth":

> For she rules over abominable (*deinos*) birth and union of all things,
> sending the female to unite with male and in opposite fashion,
> male to female. (28B12, 4–6)

The duality of male and female and their sexual attraction result in all those phenomena that the goddess expels from true being: birth, difference, plurality, growth, change, and death. The female *daimôn* of the human world is, in other words, the radical other of the goddess. It is not difficult to understand why a female goddess—and not a male god—is the "ruler" of the human world and its dependence on sexual reproduction, but why is the divine speaker of truth female? Or, perhaps first: why is the speaker of truth a divinity, instead of a mortal? Why cannot Parmenides be the subject of his own thoughts?

The answer to the last question seems obvious: the teaching of the goddess denies human experience and human knowledge about the world, as well as the basic human condition of birth and death, and as such it is in need of an outside perspective. Parmenides cannot himself be the authority of such an outside perspective, which only a non-human, or a god, can offer, not however, one of the familiar gods or goddesses of Greek mythology. These divinities do not stand outside the human world, in which they constantly intervene. By contrast, as an anonymous divinity, the speaker in Parmenides' poem is fundamentally outside of the world—of being or what is—of which she speaks the truth. But why does this speaker of truth have to be a female divinity? Why not a nameless god?

The answer to this question is, I believe, also simple. Parmenides needs a divinity that can give authority to the expulsion of phenomena for which the female sex is responsible. A goddess, far more than a god, can give him such authority. Still, it is ironic that the goddess, with the help of several other goddesses, teaches the young man a truth which expels the female from that which is. This is the truth that

[15] See Songe-Møller (2002, pp. 8–10 and 28–32).

Parmenides, through the young man's rendering of the words of the goddess, argues for and that has haunted philosophers ever since.[16]

5.3 The Role of Diotima in Plato's Symposium

Diotima's identity and her theory of love have been discussed extensively, both by feminist and non-feminist scholars. Does Plato in the *Symposium* refer to a historical woman who was a priestess in Mantinea, or does he invent this figure for the purpose of this dialogue? Most scholars today hold that "Diotima is the *one* named character Plato certainly invented" (Nails 2002, p. 137), but this has not always been so,[17] and today feminist philosophers argue convincingly for the historicity of Diotima.[18] Another question is whether there ever was a conversation between her and Socrates.[19] This we cannot determine, as we cannot determine whether the conversations in other Platonic dialogues actually took place, and even less whether Plato gives an authentic presentation of his characters and their views. Not even Socrates, who was certainly a real person, appears in the dialogues as "himself", but serves rather as a voice within Plato's complex discourse. The *Symposium* is, as other Platonic dialogues with or without historical characters, a philosophical drama, authored by Plato. My main question is accordingly: why does Plato introduce Diotima, a woman, to teach Socrates about love?

Diotima is obviously out of place in the male homoerotic gathering of the *Symposium*, a celebration of the tragic poet Agathon's victory the previous day in Athens' annual theatre festival. The topic of the evening is *Eros*, and the speeches given in praise of this god is for the most part—and not surprisingly—marked by the homoerotic atmosphere among the men as well as by the traditional Athenian paederastic ethos. In opposition to the other speakers, who have given their own view on *Eros*, Socrates does not speak in his own voice. Instead, he relates a speech (*logos*) on *Eros* which he once heard

[16]Cavarero (1995, p. 94) puts it in somewhat harsher words, when comparing the role of Diotima with that of Parmenides' goddess: 'a subtle and ambiguous strategy … ultimately reinforcing the original matricide that disinvests them'.

[17]Cf. Halperin (1999, p. 119): 'With occasional exceptions, … classical philologists of the nineteenth and early twentieth centuries tended to grant Diotima a measure of historical authenticity'. Cf. for instance Kranz (1926, p. 437): Diotima 'war zweifellos … eine historische Persönlichkeit'.

[18]See Waithe (1987), Nye (2015, pp. 1–8). Also Irigaray (1994) treats Diotima as a real person.

[19]Both Waithe (1987) and Nye (2015) claim that the conversation between Diotima and Socrates actually took place. Also Irigaray (1994) analyses the conversation between Diotima and Socrates as one that once took place. Cavarero, on the other hand, who does not exclude that Diotima was a historical person, claims that 'Diotima speaks Plato's words' (Cavarero 1995, p. 92). Kranz, who strongly argues for Diotima's historicity, claims that the dialogue between Socrates and Diotima is 'völlig freie Erfindung des Symposiondichters' (Kranz 1926, pp. 437–438).

> …from a woman of Mantinea, Diotima - a woman who was wise (*sophê*) about many things beside this: once she even put off the plague for ten years by telling the Athenians what sacrifices to make. (201d)[20]

In this short introduction, Diotima's manifold otherness is emphasized: a woman, a foreigner, and a priestess and prophetess in possession of wisdom. Also, she is absent, present only through Socrates' account of her words spoken long time ago.

Not only in his introduction of Diotima, but several times throughout his narration of their conversation, Socrates partly hints at, partly emphasizes, that Diotima was wise (*sophê*, 206b, 208b), a foreigner (*xenê*, 204c, 211d), and a prophetess (206b, 209e-210a). I take this as a clear indication that she is not pictured as a philosopher and that she speaks from another perspective than that of Socrates. However, this emphasis of her position outside philosophy creates a puzzle, if not a paradox: she is the one who develops a concept of love that fits exactly with the figure of the erotic philosopher, i.e., with Socrates himself. In the following, I will try to work out an understanding of this puzzle.

Before Socrates introduces Diotima, he interrogates Agathon on the conception of *Eros* and shows that love is a desire for something one lacks, and since love is a desire for beauty, *Eros* is not beautiful, neither is he good since what is beautiful is also good (199c–201c). This contradicts all the earlier speakers who had taken for granted that *Eros* is beautiful and thus the object of love, and not, as Socrates shows, the very subject of love. It turns out that Socrates' very Socratic interrogation and refutation of Agathon is a repetition of Diotima's refutation of Socrates' earlier misconception of *Eros*: "She used the very same arguments against me that I used against Agathon; she showed how, according to my very own speech, Love is neither beautiful nor good" (201e). Diotima continues her Socratic interrogation of Socrates—who thinks that if *Eros* is neither beautiful nor good, he must be ugly and bad—by introducing the notion of "in between" (*metaxy*): *Eros* is a *daimôn*, in between beautiful and good, ignorant and wise, mortal and immortal.

Diotima illustrates and elaborates this conception of *Eros* as something "in between" by her charming, but nevertheless startling, story about *Eros*' birth. It happened during a festival among the gods. One of the guests, the resourceful *Poros* ("way", "resource") got drunk and fell asleep in the garden of Zeus. An uninvited guest, the poor *Penia* ("poverty", "need"), turned up and "…schemed up a plan to relieve her lack of resources: she would get a child from *Poros*. So she lay beside him and got pregnant with *Eros*" (203c).

This story not only subverts the traditional roles of the sexes but even gives *Penia* and *Poros* self-contradictory natures: while the resourceful *Poros* is depicted as totally resourceless, lying drunk and asleep in the garden, *Penia* is the one who finds a way and acts as a man in her rape of the passive *Poros*.[21] The result of this unusual union is *Eros*, whose nature is a mixture of those of his parents. Although Diotima claims that it is on account of his father that *Eros* "…is brave, impetuous,

[20]The citations from Plato's *Symposium* are taken from Plato (1997).

[21]In am indebted to Oda Elisabeth Wiese Tvedt's insightful and original analysis of this myth. Cf. Tvedt (2015, pp. 74–76).

and intense, an awesome hunter, always weaving snares..." (203d), this description fits better with what we have been told about his mother.[22] Socrates' initial arguments against Agathon and the other symposiasts' understanding of *Eros* is here sharpened: whereas love in their speeches was marked by traditional Athenian sex roles and for the most part restricted to male homosexual love,[23] the story about *Eros*' birth not only has heterosexual love as its condition but also creates a confusion about the nature of femininity and masculinity.

Eros, a being in between lack (*penia*) and resource (*poros*), is nothing less than a philosopher: "*Eros* is in love with what is beautiful, and wisdom is extremely beautiful. It follows that *Eros* must be a lover of wisdom (*philosophos*) and, as such, is in between being wise and being ignorant. This, too, comes to him from his parentage, from a father who is wise and resourceful and a mother who is not wise and lacks resource" (204b). From his mother, *Eros* is poor (203c), shoeless and always in need (203d), and also his scheming nature seems to belong to his mother, although, as already mentioned, Diotima claims that this comes from his father. As Arlene Saxonhouse remarks, it is *Penia* who "gives to Eros those qualities that make him most similar to Socrates" (1995, 175). Diotima, on the other hand, is closer to *Poros*, who, like Diotima, is said to be wise (*sophos*, 204b) and therefore no philosopher.

This subversion of the sexes and undermining of traditional sex roles are further emphasized through the hierarchical relation between Diotima as the teacher and Socrates as the pupil (Sampson 2013, p. 103). Diotima even talks to Socrates as if he were a child: she laughs at his naive remarks (202b) and tells him that even a child would be able to answer Socrates' questions (204b). She leads their conversation from beginning till the end, while Socrates has little or nothing to contribute and at times has no idea what Diotima is talking about (206b).

Diotima's role has become a rather complex one. A woman, acting as a man and wise like *Poros*, gives philosophy a feminine trait, and an explicit non-philosopher gives a pertinent description of Plato's philosophical hero. By introducing an outside perspective—from a woman, a foreigner, a priestess—Plato is not only able to stir up traditional categories, but also to indicate the universality of his notion of philosophy: it is not Socrates, an Athenian citizen, talking to an intimate circle of friends, who speaks at this "originary scene of philosophy, at one of its founding moments" (Halperin 1990, p. 144), but a radical Other.

Having related the story of *Eros*' birth, Diotima undertakes a detailed, Socratic-style analysis of the concept of love—and thus of philosophy—and its mode of activity (204d-209c) and concludes that what is unique to the activity (*ergon*) of *Eros* is "giving birth in beauty, whether in body or soul" (206b). Socrates, who is by now "filled with admiration" for Diotima's wisdom, is stupefied by her words: "It would take divination (*manteia*) to figure out what you mean. I can't" (206b). Socrates is unable to understand this radical feminization of *Eros* and thus of philosophy: *Eros*' activity is said to be the one that belongs exclusively to women, namely giving birth.

---

[22] Cf. Saxonhouse (1995, p. 175).

[23] Cf. Leitao (2012, 200), who shows that the female has also been removed 'from the scene of birth in the speeches leading up to Diotima'.

Also, and perhaps above all, Socrates' astonishment at Diotima's words signals the newness of this feminized version of love and philosophy. It seems pertinent that it is presented to the group of paederasts through a perspective that was just as foreign and unheard of as this new way of thinking.

Diotima elaborates further on the feminine aspects of *Eros*: "All of us (*pantes anthropoi*) are pregnant, Socrates, both in body and soul, and, as soon as we come to a certain age, we naturally desire to give birth" (206c). Here, Diotima does not use birth as a metaphor, or at least not only as a metaphor, but gives all human beings the exclusively female capacity of being pregnant and giving birth. By this move, Diotima goes beyond beauty as the goal of *Eros*: "what *Eros* wants is not beauty, as you think it is", but "reproduction and birth in beauty" (206e). But Diotima moves further. She asks "why reproduction?" and answers: "It's because reproduction goes on forever; it is what mortals have in place of immortality" (206e-207a). By this move she has reached her final conclusion: "It follows from our argument that *Eros* must desire immortality" (207a).

So far, Diotima's conversation with Socrates can be classified as a Socratic dialogue, with questions and answers. From this point onwards, however, Diotima starts her monologue, using "language of the Mysteries" (Evans 2006, p. 5) and revealing her identity as a priestess. At this moment one can detect a certain irony from Socrates: he calls Diotima not only wise but "most wise" (*sophôtatê*, 208b) and comments that she now speaks "in the manner of a perfect sophist" (208c). Although there is here an essential break in method—from dialogical arguments to monologic teaching—there is a close thematic link between her Socratic analysis of the feminine aspects of *Eros* with her attempt to initiate Socrates into "the final and highest mysteries" (*ta telea kai epoptika*, 210a). Her claim that "all men are pregnant" can be interpreted as pointing towards the Eleusinian mysteries: according to Nancy Evans, the "particularly female experience of birth and nurturing" in the myth of Demeter and Kore are in "the Eleusinian mystical religion extended to all *anthropoi* … bestowing blessings on humans regardless of gender and human status" (Evans 2006, p. 15). It is thus possible to interpret Diotima's Socratic analysis of the notion of *eros* and philosophy as a preliminary step towards the mystical initiation of love.

Diotima refers to two stages of the initiation. She believes that Socrates might be able to be initiated into the "lesser mysteries" (208e–209e), but doubts that he is able to follow her through the "higher mysteries" (210a–212a), the well-known "ladder of love", leading from one beautiful body, via all beautiful bodies, beautiful souls, the beauty of laws and knowledge, and the "great sea of beauty" (210d), until the lover "…all of a sudden (*exaiphnês*) will catch sight of something wonderfully beautiful in its nature; that Socrates, is the reason for all his earlier labors" (210e–211a). At this point, the lover is in touch with beauty itself, which "…always *is* and neither comes to be nor passes away, neither waxes or wanes…" (211a), where he gives birth "not to images of virtue … but to true virtue." Here, Diotima utters her last words: "If any human being (*anthropos*) would become immortal, it would be he" (212a).

It is not dialogical thinking that has led the lover up the ladder, but "correctly" (*orthôs*) following "the right order" (210e) and "through the correct paederasty" (211b). These last words might surprise some readers: they do not fit nicely with the feminization of *eros*, nor with the fact that Diotima, for the most part, has talked about the lover as an *anthropos* and also has included heterosexual love in her discourse. Diotima's speech is in this respect ambiguous. How should we understand her discourse? Does it, as my interpretation indicates, represent a break with the homoerotic environment of which her speech forms the central part, as well as with the notion of the subject of philosophy as male, as the Platonic dialogues otherwise substantiate? Or is Diotima's speech, in reality, nothing but a male strategy of expropriation, transferring the female capacity of giving birth to the male and thus rendering woman superfluous? These questions have been extensively discussed, and the views vary.[24] I believe the text itself is deliberately ambiguous, which should not surprise us, since Diotima's teaching must have appeared both strange and upsetting to the group of Athenian citizens and paederasts, but which is somewhat softened through her several examples of paederastic love throughout her speech (for instance 209c, 210c) and which opens for an interpretation in another direction than the one I have given here.[25]

What, then, is Diotima's function in this last part of her speech, where she describes an initiation into "the final and highest mysteries"? The lover has here reached the region of true being (212a), which is beyond the reach of the *daimôn Eros*, the imperfect son of *Penia* and *Poros*, i.e., beyond the reach of the Socratic philosopher, who was defined through his lack of knowledge. It is quite consistent that Diotima in this latter part of her speech leaves her dialogical method and starts to speak as a priestess, or a mystagogue: in order to indicate the goal of Plato's philosophy, which is truth and which, perhaps paradoxically, transcends his own definition of philosophy, Plato needs a person who has a privileged access to the divine. It could have been an Athenian priest, but a foreign priestess underscores that the perspective is that of the Other.

5.4 Conclusion

Is there any parallel between the function of Diotima in the *Symposium* and that of the goddess in Parmenides' poem? One parallel is striking: both Diotima and the goddess tell a truth that is beyond the reach of ordinary human knowledge. For both of

[24]See for instance Saxonhouse (1995) and Nye (2015), who argue for a heterosexual interpretation of *eros*, and Tvedt (2015, pp. 77–79), who interprets *eros* as a melding of male and female. On the other hand, Cavarero (1995), Du Bois (1988), and Halperin (1990) argue for a male appropriation of the female in Diotima's speech. According to Irigaray (1994), Diotima in the first part of her speech speaks in her own, feminine voice. At a certain point, however, 'Diotima's method miscarries', perhaps distorted by Socrates (188). Hobbs (2006, p. 271) argues that 'Plato is chiefly concerned not with "appropriating the feminine" but with liberating men and women alike from inessential bodily and cultural constraints.' These are just a few examples from the extent literature on this topic.

[25]In Songe-Møller (2002, Chap. 5), I take this other direction, which, however, I now believe is somewhat one-sided.

them, true being is characterized by the absence of difference, plurality, and change, which gives access to immortality. And for both of them, immortality is the final goal of their teaching. Both Parmenides and Plato make use of female characters—a goddess and a priestess—in order to construct a non-human perspective on being, a perspective of the Other.

There are, however, radical differences between the teaching of the goddess and Diotima. While Parmenides' goddess teaches that birth, and also *eros*, leads to death and therefore must be expelled from true thinking and being, Diotima teaches the opposite: *eros* and its activity, which is birth, is the only thing that can lead a human to intimate company with true being and thus to immortality. While Parmenides uses a female divine character to expel all the phenomena for which the female sex is responsible, Plato does the opposite: he uses a female character to include what is exclusively female into his philosophy. For Diotima, philosophy *is* birth and, as such, creativity.

While Parmenides in the first part of his poem seems to use non-human female characters in order to emphasize that the quest for knowledge belongs to man alone, Plato has a radically different agenda: in the first part of Diotima's speech, both the traditional notions of male and female and the subject of philosophy are destabilized. Diotima—a woman, a foreigner, and a priestess—enabled Plato to do this. One could take this a step further: it took a woman to point towards a non-gendered subject of philosophy.

It is remarkable that two female characters—the goddess and Diotima—play such decisive roles in these two texts, which may be said—each in its way—to have determined much of the course of Western philosophy. Parmenides uses the goddess to expel everything female, while Plato uses Diotima to include the female into his philosophy. In each case, the female is *used* by male philosophers. In each case, the female represents the Other, which enables Parmenides and Plato to create an outside perspective, each in their own way.

References

Arthur, M. B. (1983). The Dream of a World without Women: Poetics and the Circles of Order in the *Theogony* Prooemium. *Arethusa, 16,* 97–116.

Canters, H., & Jantzen, G. M. (2005). *Forever Fluid. A Reading of Luce Irigaray's Elemental Passions*. Manchester/New York: Manchester University Press.

Cavarero, A. (1995). Diotima. In *In spite of Plato*, (pp. 91–120), (S. Anderlini-D'Onofrio & Á. O'Healy, trans.). New York: Routledge.

Coxon, A. H. (2009). *The Fragments of Parmenides. A Critical Text with Introduction and Translation, the Ancient Testimonia and a Commentary*. Las Vegas/Zurich/Athens: Parmenides Publishing.

Curd, P. (2011). *A Presocratics Reader. Selected Fragments and Testimonia. Second Edition*. Ed., with introduction, by P. Curd (R. D. McKirahan & P. Curd trans). Indianapolis/Cambridge: Hackett Publishing Company.

Diels, Hermann/Kranz, Walther. (1951). *Die Fragmente der Vorsokratiker. Griechisch und deutsch. Erster Band*. Berlin: Weidmann.

Du Bois, P. (1988). The platonic appropriation of reproduction. In *Sowing the Body: Psychoanalysis and Ancient Representations of Women* (pp. 169–183). Chicago: University of Chicago Press.
Evans, N. (2006). Diotima and Demeter as Mystagogues in Plato's *Symposium*. *Hypatia 21*(2), 1–27.
Halperin, D. (1990). Why is Diotima a woman? In *One Hundred Years of Homosexuality* (pp. 113–151). New York/London: Routledge.
Heidegger, M. (1971). *Was heisst Denken?*. Tübingen: Max Niemeyer Verlag.
Henry, M. M. (1995). *Prisoner of History. Aspasia of Miletus and Her Biographical Tradition*. New York/Oxford: Oxford University Press.
Hesiod. (1966). *Theogony*. Ed. with Prolegomena and Commentary by M. L. West. Oxford: Clarendon Press.
Hobbs, A. (2006). Female imagery in plato. In *Plato's Symposium. Issues in Interpretation and Reception*, ed. J. H. Lesher, D. Nails, & F. C.C. Sheffield (pp. 252–271). Washington, DC: Center for Hellenic Studies.
Irigaray, L. (1994). Sorcerer love: A Reading of Plato's *Symposium*, Diotima's Speech. (E. H. Kuykendall trans.). In *Feminist Interpretations of Plato*, ed. Nancy Tuana (pp. 181–195). University Park: Pennsylvania State University Press.
Kranz, W. (1926). Diotima von Mantinea. *Hermes, 61*(4), 437–447.
Leitao, D. D. (2012). The pregnant philosopher. Masculine and feminine procreative styles in Plato's *Symposium*. In *The Pregnant Male as Myth and Metaphor in Classical Greek Literature* (pp. 188–226). Cambridge: Cambridge University Press.
Loraux, N. (2006). *The Invention of Athens. The Funeral Oration in the Classical City* (A. Sheridan, trans.) New York: Zone Books.
Nails, D. (2002). *The People of Plato. A Prosography of Plato and Other Socratics*. Indianapolis/Cambridge: Hackett Publishing Company.
Nye, A. (1990). *Words of Power. A Feminist Reading of the History of Logic*. New York/London: Routledge.
Nye, A. (2015). *Socrates and Diotima. Sexuality, Religion, and the Nature of Divinity*. New York: Palgrave Macmillan.
Plato. (1997). *Symposium*. In J. Cooper (ed.), (A. Nehamas & P. Woodruff trans.). *Plato. Complete Works*. Indianapolis/Cambridge: Hackett Publishing Company.
Sampson, K. (2013). The philosophical significance of the figure of Diotima. *Norsk filosofisk tidsskrift, 48,* 100–109.
Saxonhouse, A. (1995). Plato's *Smposium*: A reassessment of Callipolis. In *Fear of Diversity. The Birth of Political Science and Greek Thought*. Chicago/London: The University of Chicago Press.
Songe-Møller, V. (2002). *Philosophy Without Women. The Birth of Sexism in Western Thought*. London/New York: Continuum.
Tarán, L. (1965). *Parmenides. A Text with Translation, Commentary, and Critical Essays*. Princeton/New Jersey: Princeton University Press.
Tvedt, O. E. W. (2015). *Fødselens ontologi. Kjønnsforskjellen i Platons og Batailles tenkning om det erotiske begjær* (*The Ontology of Birth. Gender Difference in Plato's and Bataille's Philosophy of Erotic Desire*). Master thesis, University of Bergen.
Vernant, J.-P. (1983). *Myth and Thought Among the Greeks*. London/Baltmore/ Melbourne/ Henley: Routledge & Keagan Paul.
Vidal-Naquet, P. (1986). *The Black Hunter* (A. Szegedy-Maszak trans.). Baltimore: Johns Hopkins University Press.
Waithe, M. E. (1987). Diotima of Mantinea. In *A History of Women Philosophers. Volume 1/600 BC—500 AD*, ed. Mary Ellen Waithe, 83–116. Dordrecht/Boston/ Lancaster: Martinus Nijhoff.

Chapter 6
The Torn Robe of Philosophy: Philosophy as a Woman in *The Consolation of Philosophy* by Boethius

Sigridur Thorgeirsdottir

Abstract Symbolic figures like Sophia, Philosophia or Lady Reason represent feminine features in texts of the Western philosophical tradition that are often overlooked in their later interpretations. *The Consolation of Philosophy* by Boethius (480–524), one of the most widely read philosophical texts of medieval times, includes a dialogue between the imprisoned Boethius who awaits his death sentence and Philosophia, a feminine personification of philosophy. In my interpretation of Philosophia, I analyze how the practice of philosophy she and Boethius stage in this text consists of working with and reflecting on the difficult emotions he struggles with. This argument is based on how ancient meanings of the noun *sophia* include practical, embodied, and sensual knowledge and not only theoretical knowledge. My interpretation hence involves underscoring feminine elements of philosophical reasoning that includes embodiment and emotions. Philosophia resurfaces in many philosophical texts, such as in Christine de Pizan's *The Book of the City of Ladies* (1405), one of the greatest feminist books of the middle ages, where Lady Reason teaches the author to help her trust her feelings and judgements about women.

6.1 Introduction

When I was in my first year of studying philosophy, my mother gave me a book. It was one of the classics of medieval philosophy, *The Consolation of Philosophy* by Boethius (480–524). The title did not appeal to me at the time. The concept of consolation sounded to me too theological. I wanted to avoid that since a course in theology during my gap year before the university had attracted me and yet turned me off at the same time. Truth in theology seemed to be consoling truth, and my young self felt it needed more questioning and critique of truth rather than consolation. Hence philosophy came along with its open attitude of critical questioning of supposedly self-evident truths. At the beginning of my life in philosophy *The Consolation of*

S. Thorgeirsdottir (✉)
Department of Philosophy, University of Iceland, 101 Reykjavik, Iceland
e-mail: sigrthor@hi.is

S. Thorgeirsdottir and R. E. Hagengruber (eds.), *Methodological Reflections on Women's Contribution and Influence in the History of Philosophy*, Women in the History of Philosophy and Sciences 3, https://doi.org/10.1007/978-3-030-44421-1_6

Philosophy did not appear to me as it did after years and years of familiarizing myself with the Western tradition of philosophy. I was struck when I recently picked up this book that had spent silent decades in my bookshelf and expedition boxes. This is a dramatic book, with a slight resemblance to the *Phaedo* about Socrates' last days. As it opens, the philosopher Boethius, who had a high office as consul and political advisor, has been imprisoned for political reasons and awaits his sentence and execution. Unlike Socrates, Boethius did not have a chance to defend himself at a trial, and he is alone in prison and not accompanied by his friends or loved ones in this dark hour. In despair, this man who was also one of the great philosophers of his times turns to philosophy for consolation. He begins the book by describing how philosophy appears to him as Philosophy, the lady of philosophical wisdom[1]:

> While I was quietly thinking these thoughts over to myself and giving vent to my sorrow with the help of my pen, I became aware of a woman standing over me. She was of awe-inspiring appearance, her eyes burning and keen beyond the usual power of men. (Boethius 1983, 35)[2]

This is a dramatic opening, and the whole book is, in fact, a theatrical performance of a dialogue between Boethius and Philosophy, mixed with monologues and songs of her. What first struck me about the figure of Philosophy and the book when I finally got to read it was that the woman Philosophy, reminded me in some way of Diotima in Plato's *Symposium*. Diotima is a woman and at the same time, she is an elevated figure that is described as wise (*sophe*) and having achieved wisdom in contrast to the wisdom lovers in the *Symposium*. There is the aura of unreality about Diotima that has caused a never-ending dispute among scholars about if Diotima really existed or not. Contemporary feminist philosophers like Nye and Waithe claim that Diotima was a real woman who existed and was a teacher of Socrates (Nye 1989; Waithe 1987). In their interpretations, Diotima moreover represents feminine wisdom that goes against the male-centric thrust of the text.

Obviously, there cannot be a debate about Philosophy as a real historical person. She is a symbolic figure of philosophy that is at times personalized as a Sophia figure but also often depersonalized as a symbol of wisdom. In *The Consolation of Philosophy* she is understood to be a "supernatural entity" (Shanzer 2009, 231). Philosophy has many womanly roles that range from "Athena-like divine epiphany", "impatient or jealous mistress", "kind mother and goddess", "doctor" to "former nurse" (Shanzer 2009, 232). The woman who speaks, recites, sings, converses, performs, and acts out these roles in the dialogue with Boethius is in most interpretations of the text void of important womanly traits. When I speak about "womanly" traits I do not mean her womanly roles in the text but rather features that are traditionally associated with women in the history of Western philosophy, features that have to do with embodiment and emotions. There are several references in the text to her as a woman (lat. *mulier*) and that alone gives reason for attending to the possible meaning of the womanhood of Philosophy in light of the philosophical context and background of

[1] In the English translation of the text the woman is referred to as Philosophy, but in the Latin version she is Philosophia, which displays more clearly the connection with the ancient Greek noun, *sophia* (wisdom).

[2] Boethius. 1983. *The consolation of philosophy*. Trans. by V.E. Watts. London: Penguin.

the text.[3] The lack of attention to womanly embodiment in the figure of Philosophy rendering her as a neuter figure has to do with the influence of Platonic and Christian thought. Like Shanzer points out in her discussion of Philosophy, Augustine "experienced growing anxieties about figures" like Philosophy after his conversion, "because they seemed to be pagan holdovers" (Shanzer 2009, 231). Shanzer hence comes to the conclusion that after Boethius "the fate of such personified learned ladies was secure—they were there to stay and became domesticated goddesses in the Middle Ages" (ibid., 232). Unlike Augustine, Boethius had no "qualms about taking over an unabashedly pagan form of encounter without bothering to Christianize it" (ibid., 232). For my interpretation of Philosophy, it is important to view her as representing a tension between pagan and Christian outlooks. Disembodied interpretations of Sophia on the other hand only display certain features of her and conceal others, hence disregarding her as a principle of integrated wisdom that combines theoretical and practical wisdom, thinking and feeling.

The Christian-Augustinian view of Philosophy has roots in the ancient Greek enlightenment, i.e., in Plato's philosophy. The dichotomy of body and soul had the woman associated with the body and man with the soul. The body, moreover, became associated with impurity, transience, and lack of rational control. The view of the figure of Philosophy as a supernatural, feminine entity hence happens via a rejection of the meaning of her embodied features. The notion of the word *sophia* is of importance here. As Tsakiridou argues in her study of the ancient Greek meaning of the noun *sophia*, the term shifts during the ancient Greek enlightenment towards masculine activities and only formally retains the feminine delineation (Tsakiridou 1999, 239). The noun *sophia* was formerly associated with sensual knowledge like Tsakiridou points out in her study of the gender of ancient Greek words that have to do with womanhood. Furthermore, *sophia* involves the plurisensual that includes the "tactile, visual, audial, kinetic and affective/somatic states" (Tsakiridou 1999, 260). The etymological roots of *sophia* suggest, according to Tsakiridou, original meaning of sensual contact between the knower and the known (Tsakiridou 1999, 239). According to Liddell and Scott's Lexicon, *sophia* means "cleverness or skill in handcraft and art" (Liddel and Scott 1996, 1621). The noun *sophia* hence originally meant a kind of practical wisdom, in comparison to Plato's and Aristotle's understanding of *sophia* as theoretical wisdom.

Hannah Arendt who discusses *The Consolation of Philosophy* in her book on *The Life of the Mind,* draws attention to how an "unawareness of the body" in the thinking experience is apparent in ancient Greek philosophy. "Thinking," writes Arendt, "implies an unawareness of the body and the self and puts in their place the experience of sheer activity more gratifying, according to Aristotle, than the satisfaction of all the other desires" (Arendt 1978, 162). This idea explains, in Arendt's view, the "curiously extreme theories of the power of mind over body—theories clearly refuted by common experience" (Arendt 1978, 163). She quotes Gibbons who pointed out how such an idea of consoling thinking is "ineffectual to subdue the feelings of human

[3]For the original text in Latin see: http://faculty.georgetown.edu/jod/boethius/jkok/boeconc/main.htm.

nature" (Arendt 1978, 163). Yet this "strong hostility to the body … [was] adopted virtually intact by the Christian creed" (Arendt 1978, 163).

The text of Boethius is in most interpretations of it seen to adhere to a strict duality of body and soul. Despite this overall direction of ancient Greek philosophy, a closer look at Philosophy in *The Consolation of Philosophy* gives us a richer picture of her. When one opens one's eyes to ancient meaning of *sophia* as wisdom, Philosophy appears as a much more complex and multifaceted figure than most interpretations of Boethius' text acknowledge. She represents wisdom as wise judgment that involves reason and emotion. In the following interpretation that I offer, I emphasize how Philosophy in Boethius' text is a figure that conveys sexual difference and embodiment. Lady Philosophy represents, as I will show, that the human being is a composite of body and soul, reason and emotion. Attention to these features of her shed fresh light on the idea of philosophical thinking Philosophy practices in her dialogues with Boethius.

6.2 The No-Place of Woman in Philosophy

In the introduction to *The Consolation of Philosophy* by V.E. Watts the figure of Philosophy is made vague and opaque: "The whole [of Boethius' text]… is cast in the form of a particular type of dialogue, the sacred dialogue, which the author describes how *some* divine spirit or power, at first unknown to him, appears and reveals to him *some* portion of hidden wisdom" (Boethius 1983, 19). The italics are mine, drawing attention to how Philosophy is made illusive as "some" divine spirit who reveals "some portion of hidden wisdom". Philosophy is made less visible and less significant with this depiction. This opaqueness of Philosophy in interpretations like Watts' is an indication of the no-place of women in Western philosophy that Luce Irgaray has discussed in her readings of classical texts of the Western philosophical canon. Sexual difference as a core dualistic distinction within philosophy of our tradition has in Irigaray's view prevented the flourishing of what she calls a "female subject position" (Irigaray 1985, 1993).

Irigaray's idea of the no-place of women is new to feminist scholarship in philosophy in the last decades of the twentieth century. According to Beauvoir's philosophy of the *Second Sex* women have been allocated the position of the second sex, a sex that is secondary and inferior to the first and absolute sex which is the male sex (Beauvoir 2009). Women have been defined negatively and hence Beauvoir's philosophy is directed at Aristotle's philosophy of the female sex defined as lacking features of the superior male sex that laid the ground for traditional Western philosophical misogyny. Beauvoir's view of the feminine in the history of philosophy has been a mainstream view in contemporary feminist philosophy. Irigaray, on the other hand, maintains that women in Western philosophy have not even had a position of an other sex insofar they have only been an other to the male sex, an other of the same; a sex in which the male sex mirrors itself to elevate itself as she argues in her book *Speculum of the Other Woman* (1985). Irigaray hence goes further than Beauvoir with her idea of how the female sex as the second sex has been defined in the history of philosophy. The no-place of a female subject-position in philosophy, Irigaray points out, becomes apparent in the silencing and erasing of the woman or the feminine as happens with Philosophy in interpretations of Boethius' text. Yet these efforts leave traces as well as hidden womanly features and empty spaces. What happens to Philosophy in interpretations of her can be understood in this light. She is made opaque as a woman and given neuter attributes. Such interpretations happen in the mainstream Platonic context that provides the background for gender-neutralizing philosophical interpretations of the meaning of the word *sophia*. In *Theatetus* (145e) Plato associates *sophia* with abstract, speculative knowledge of the forms, and in Aristotle's *Metaphysics sophia* is identified as "first philosophy".

6.3 The Torn Robe and the History of Philosophy

Canonical and marginal texts in philosophy require scholarship that attends to overt and latent meanings and interpretations of names and terms like Philosophy and *sophia*; names and terms that have womanly features or connotations. The text by Boethius is an example of the need to go beyond the alternative views of Philosophy as either representing abstract, speculative metaphysical knowledge or as seeing her as a relict of a mythological, pagan, ancient Greek culture. There is no denying that the philosophy of Boethius stands very much in the Platonic tradition in terms of topics, form, and content. Philosophy describes herself at the beginning of the text very much as the philosophical wisdom represented by Plato and Socrates. After the unjust death of Socrates "the mobs Epicureans and Stoics … did all they could to seize for themselves. They, so Philosophy:

> tried to seize for themselves the inheritance of wisdom that he [Socrates] left. As part of their plunder they tried to carry me off, but I fought and struggled, and in the fight the robe was torn which I had woven with my own hands. They tore off little pieces from it and went away in the fond belief that they had obtained the whole of philosophy. The sight of traces of my clothing on them gained them the reputation among the ignorant of being my familiars, and as a result many of them became corrupted by the ignorance of the uninitiated mob. (Boethius 1983, 39)

Philosophy hence divides fake philosophers from the real ones who were persecuted, tortured or harassed because of their contempt of the pursuits of immoral men that her teaching had instilled in them (Boethius 1983, 39).

Philosophy's description of the tearing of her robe is quite revealing. These fake philosophers attack her, undress her and tear her robe. There is violence in this image. This scene reminds of the much more brutal attack on Hypatia (370–415) who was killed by a warring mob in Alexandria, and her body torn to pieces. In fact, the murder of Hypatia, one of the foremost scientists of late antiquity, is a bloody sacrifice of a womanly authority that would continue and culminate in the witch hunts at the beginning of the age of modern science.

The torn robe of philosophical wisdom is a rich and multifaceted image that is important for the study of women and feminine figures in the history of philosophy. Philosophy has, like she states, woven the robe herself, and the philosophers whom she and Boethius do not approve of adorn themselves with pieces of the robe. From the standpoint of research into women in the history of philosophy—when we include symbolic figures of womanhood and femininity like Philosophy—the image of the torn robe illustrates the fate of women and the feminine in the philosophical tradition. Writings of women philosophers have throughout the ages, if they did not get lost or "forgotten", sometimes been published under male names. Discoveries of women philosophers that have been influential in their respective historical settings and contexts have for the most part been disregarded and excluded when it comes to writing the histories of these philosophical periods and schools. Encyclopedias of philosophy and philosophers have until recently hardly mentioned women philosophers. Making Philosophy a more or less genderless symbol of wisdom as is the case in most interpretations of Boethius's text amounts to silencing of her feminine features.

6.4 Christine de Pizan and Lady Reason

The silencing of womanly features of Philosophy in Boethius' book is all the more significant given the fact that the most notable rendering of this figure is to be found in Christine de Pizan's *The Book of The City of Ladies* that came out in 1405, and is considered one of the greatest feminist books of the middle ages (Pizan 1999). Christine de Pizan was one of the first professional writers, a single mother who earned her living with writing. The structure of the book is in important respects taken over from Boethius's work. The figure of Philosophy is introduced as Lady Reason, and like *The Consolation of Philosophy* by Boethius, *The Book of the City of Ladies* begins with the lament of the author, Christine de Pizan herself. "Sunk in … unhappy thoughts, my head bowed as if in shame and my eyes full of tears" (Pizan 1999, 7). She feels desolate after reading a work from the thirteenth century by an author by the name of Matheolus who in his reflections on marriage writes that women make men's lives miserable. This was but one author of many that Christine de Pizan read who expressed his contempt for women. She ponders on these depictions of vicious women and compares to her own experiences and to women from different classes and walks of life. Even though her introspection does not give any support to the shaming of women, she has a hard time figuring out why so many books that are supposed to be constructive nevertheless contain such tirades against women (Thorgeirsdottir 2020). It is hard to believe that so many wise and learned men were so united in lying about women. So, Christine feels deeply insecure because she has kind of unlearned how to trust her own feelings and her own judgement. At this moment of desperation about herself and the nature of her sex, she has a vision of how Philosophy as Lady Reason appears to help her to relearn to trust her own feelings and judgement. Lady Reason is accompanied by two other

Ladies, Lady Rectitude and Lady Justice. They are there to help her to learn the truth about women, and also about the men who are so full of contempt for them. Lady Rectitude represents logical thinking and Lady Justice represents moral conduct. Lady Reason is the foremost figure of the three and has a key role in this book about how the three of them build an allegorical *City of Ladies*. It is an allegorical city–also similar to Augustine's book *City of God*–in which knowledge is refashioned by uprooting misogyny. The walls of the city that is built on new knowledge protect women from the slander and lies about them. The book, therefore, offers a history of misogyny and counters it with descriptions of great and dignified women of the past, historical figures as well as allegorical figures. Remembering these great women helps Christine de Pizan gain confidence in herself and in her sex. The figure of Lady Reason helps her reflect and reason about the virtues and talents of women and assists her in learning to trust her own judgement. "Return to your senses and stop worrying your head about such foolishness" Lady Reason tells her (Pizan 1999, 8). She similarly argues that men who hate women go against reason and against nature.

There are numerous references to nature in the book. It is nature that wills that women are worthy of protection and celebration. It is nature that decides where the City of Ladies shall be situated. Lady Reason also tells Christine to listen to her own self, and not to go against her nature in her judgement. So an embodied notion of integrated knowledge is also visible in Lady Reason's references to nature.

6.5 Reason and Emotion: From a Disembodied to an Embodied Philosophy

Silencing and denying the embodied features of Philosophy by restricting her to abstract, speculative wisdom makes us overlook subtler features of the emotional aspects of the dialogue between Boethius and Philosophy. The Christian view of Sophia tends in the same direction as the Platonic tradition by elevating Sophia to a figure that represents Christian wisdom and worldview. She is a lady of Christianity, identified with the Christian logos. However, with the development of Christian theology in the middle ages interpreters like the nineth century monk Bovo of Corvey began to point out pagan elements in *The Consolation of Philosophy* not acknowledged in interpretations of it inspired by the Platonic-Christian approaches to it (Nash-Marshall 2012, 178). Given the historical setting of Boethius in a time of transition between and pagan and Platonic-Christian worldviews, it was quite self-evident to point out that there were elements in *The Consolation of Philosophy* that were contrary to the mainstream Christian doctrines. Yet, contemporary scholars specialized in the transitional, historical context of Boethius still hardly discuss ancient sensory and affective meanings of *sophia* in the text, for example, evident in the collection of articles in *The Companion to Boethius in the Middle Ages* (Kaylor and Phillips 2012).

Sophia emerges as a goddess in the form of lady of wisdom in the gnostic tradition of early Christianity (in *Pistis Sophia*). There are exceptions to the non-acknowledgement of Sophia in the Middle Ages and early modern philosophy, especially in philosophies of mysticism. For Hildegard von Bingen, Sophia was a cosmic figure. The seventeenth century Christian mystic, Jane Leade, wrote about visions and dialogues with "Virgin Sophia" who revealed spiritual workings of the universe very much in the spirit of the German mystic Jakob Boehme (Smith 1977). Carl Gustav Jung considered Sophia to be an important figure that we need to remember in times of dissolutions of traditional ideas about God (Jung 2009). God remains incomplete without a female counterpart in Jung's view that has been there all along but overlooked. The *Old Testament* "Proverbs" (8:22–31) mentions how Sophia, mostly translated as "wisdom", is at God's side during creation.

Lady Philosophy and Sophia are not the same figures, but they certainly overlap in Boethius' text where there is an intertwining of ancient Greek and Christian dimensions. Like the above examples indicate, there are more diversified views on Sophia to be found that go beyond the defeminized and depersonalized image of Philosophy as a symbol of speculative wisdom in philosophy. It is, however, not until with recent feminist philosophical scholarship that a richer image of Philosophy in the text of Boethius emerges. This is a scholarship that accounts for the fact that Philosophy introduces sexual difference and embodiment into the text. That does not amount to a regression into a pagan outlook by reverting to the framework of the duality of body and soul, nature and culture. On the contrary, Annegret Stopczyk-Pfundstein's book on Sophia's body (*Sophias Leib*) contains a short section on Philosophy's role in *The Consolation of Philosophy* where she gives special attention to how Philosophy as a Sophia figure teaches Boethius not to divide body and soul but to strive for embodied wisdom (Stopczyk-Pfundstein 2003, 222). Stopczyk-Pfundstein writes that Sophia advises Boethius to "sink into his own body, to ignite in this way an 'inner brightness' or wisdom that will become his own" (Stopczyk-Pfundstein 2003, 222, my translation). This approach has been an important point of departure for this interpretation of Philosophy in *The Consolation of Philosophy*. Sophia displays a practice of embodied wisdom and deepening of philosophical thinking and judgment. Embodied philosophical thinking is emotion-laden. The imprisoned and condemned Boethius is in despair. In the opening scene of the book, he is crying (Boethius 1983, 36). He philosophizes out of a difficult emotion that he tries to come to terms with, with the help of Sophia. In these efforts, he comes to reflect on classical questions of philosophy.

All thinking is embodied insofar the mind is embodied (Schoeller and Thorgeirsdottir 2019). Thought and emotions are intertwined because cognition is intentional and affections have a cognitive component. This fact that is amply researched within present day cognitive science accounted for in philosophies of emotions is, however, not what is at stake here. In *The Consolation of Philosophy,* we have an example of how in philosophical reflection emotions and sentiments are acknowledged as part of the philosophical thinking process. It is our body that keeps the memory of all that we live and have experienced, our knowledge, wisdom, and sense of things. Our felt sense of things and our emotive grasp of issues are a resource for creative and intuitive thinking. The methodology I am adopting for this interpretation consists of uncovering traces of elements of Philosophy as a Sophia figure that are neglected in conventional, traditional readings of *The Consolation of Philosophy*. These have to do with philosophical thinking as affective thinking in the sense of acknowledging the desires and the fears that accompany thinking. We as philosophers are embodied thinking beings who philosophize not with a "view from nowhere" but from a given position, perspective and affective disposition (Nagel 1989). We also think with our hearts. The story of Boethius and Philosophy in the prison is about him talking out of despair and her talking with him and singing to him in an effort to help him come to terms with his predicament. Boethius is philosophizing out of a desperate position, but it is also just an extreme form of a basic affective position as a point of departure for any philosophical thinking process.

6.6 Descending into the Body

At the beginning of their discussions, Philosophy says to Boethius "you have forgotten your true nature" (Boethius 1983, 51). She reminds him that the seed of truth lies deep within him. The philosopher as a seeker of truth must hence turn his "inward gaze" unto himself for his heart holds the truth "in its own treasure chests within" (Boethius 1983, 108). These passages do not only affirm the Platonic idea of knowledge from the *Meno* as remembering knowledge stored in the soul. They deepen this Platonic insight by taking embodiment into account for these truths are, according to Philosophy, "hidden in the body's density" (Boethius 1983, 154).

Even though Boethius adheres in his philosophy to Platonic dualism of body and soul he does not strive for repression of emotions. His story in the prison-dialogues with Philosophy is neither about blocking off emotions by being thick-skinned nor being totally overwhelmed by them, but rather relating to them. By relating to his own emotions, he can get a deeper sense of the philosophical questions that are raised in a dialogue that has the goal of consoling him. So Philosophy addresses his emotions and aids him in working with them. The very fact that she sings, i.e. uses music is an indication of this because music is able to affect us in a deep, profound and embodied way. Philosophy says that music is the "maid servant"of her house. Her music is like mild and pleasant nourishment that Boethius' body absorbs to prepare it for her teaching (Boethius 1983, 54). Yet Philosophy makes one clear distinction right at the outset. Her singing is not to be confused with the song of the muses of poetry she shames as sirens and "hysterical" because they manipulate men with their passionate singing, making them go out of their minds (Boethius 1983, 36). Her therapy of the emotions is milder and subtler as the music is supposed to sing him melodies of "varying moods" (Boethius 1983, 54). Yet the music is only a preparation, for Philosophy admits that once it stops sounding in his ears, "the mind is weighed down again by its deep-seated melancholy" (Boethius 1983, 59). What is therefore needed after this attunement to the emotions is something that is calculated to "penetrate deep inside" (Boethius 1983, 59). Philosophy trains Boethius in going deeper into sensations of his own body.

Going inward and deep down is in line with the concept of *sophia* as practical knowledge of the embodied and sensing human being. In the case of Boethius this entails not only attentive listening in a conversation but also just as importantly a deep listening to himself; connecting with and attuning to his sentiments. That relieves emotional tension and gives comfort to his exhausted spirit. Working with the passions heals their disturbing effects. "It is 'says Philosophy', as if you had become swollen and calloused under the influence of these disturbing passions, and by their more gentle action they will temper you ready to receive the strength of a sharper medicament" (Boethius 1983, 49).

We should not restrict the dialogue of Boethius and Philosophy to a therapeutic endeavor with the sole goal of consoling one who is awaiting a death sentence. The consoling involves philosophical questioning of truths that are thought to be self-evident or in need of scrutiny and differentiation. That requires dialectical and logical thinking that goes hand in hand with a reflection of deeply-held values and sources of meanings and beliefs. This method is meant to liberate Boethius from his melancholy, and that has to go hand in hand with a deepening of his understanding of meanings of concepts he ponders on. This kind of thinking enables him to think something that, deep inside, he knows, i.e. to recollect it. It requires connecting with the "touchwood" that burns "deep in the heart" (Boethius 1983, 109). This connection with the body is not to be confused with being "imprisoned in earthly flesh" in the sense of being driven by passions that hinder philosophical thinking (Boethius 1983, 149). It is rather a method of meditative attentiveness, "silent attention", and clarification of the concepts at hand (Boethius 1983, 78). Philosophy compares the clarification in preparing the soil for cultivation. If one wants to sow into virgin soil, one first has to free the fields of bush and brambles with the scythe (Boethius 1983, 79). One has to clarify philosophical concepts and principles to be able to see them in a fresh way on the basis of the embodied, experiential wisdom one has.

6.7 Philosophy and Sexual Difference

In addition to displaying the body as a composite of body and soul, Philosophy also represents a sexual difference in the text. And that is the second important feature of her that I want to draw attention to, one that is mostly overlooked in interpretations of her meaning in the text. She reminds Boethius of his origin in his mother's womb, and she reminds him of the maternal energy that resides within him as she describes herself as a maternal origin of his wisdom and knowledge (Boethius 1983, 56). Her role in the text resembles in this regard the goddess who teaches Parmenides and introduces him to her wisdom. The connection to the goddess is what makes these philosophers real, and not just like the ones that adorn themselves with pieces of her robe.

The feminine element Philosophy introduces has to do with the sexuate (sex and situatedness) body. Sexual difference is a profound difference in nature-culture with infinite variations and possibilities of becoming. The difference of the sexes introduces a space between them where they affect and transform each other through the encounter of their differences. Tsakiridou argues that the sensual contact between the knower and the known the noun *sophia* represents is probably, but not exclusively "based on sexual union and reproduction" (Tsakiridou 1999, 239). It does not take much psychological imagination to see the sexual metaphorical connotation in the first paragraph of the book where Boethius writes that as he is thinking and giving went to his sorrow "with the help of my pen" he becomes aware of a woman standing over him (Boethius 1983, 35). Such dimensions have traditionally been rejected. In the fourteenth century interpretation of Boethius' text by Conrad Humery there are for example explicit efforts to deny such erotic and sensual dimensions. The character of the "Boethius-Philosophy" relation is turned by Humery into a "cordial friendship" like Hehle argues (Hehle 2012, 308).

In epistemological terms, the idea of sensual contact is a model for philosophical thinking that displays how thinking begins with the examination of the senses and sensations. The embodied, feminine Philosophy is in my interpretation a model for sensual knowledge. For Tsakiridou sophical knowledge consists in the ability such as "being able to tell whether a child could be conceived, under what conditions, and whether it would be healthy, and then by analogy, whether a fruit was ripe for collection, whether a field would yield a good crop, and so on (Tsakiridou 1999, 239). Such knowledge requires an experienced sense, information, understanding as well as intuition and the ability to make sound judgments. Another insightful connoisseur of ancient Greek culture and thought, Nietzsche, points out that the meaning of the word *sophia* is "taste". The wise man, the philosopher, is accordingly the "man of taste".[4] I, therefore, argue that these archaic aspects of *sophia* as taste, as sensual knowledge are present in the conception of philosophical thinking that emerges through the dialogue of Philosophy and Boethius. Taste is here not to be understood as a mere judgment about the beauty of an artwork. That is a notion of taste that comes up with the emergence of modern aesthetics as a subdiscipline of philosophy. Taste is really here in *The Consolation of Philosophy* a sign of philosophical judgment. To be wise means the ability to taste and discern and make a judgment.[5] This is also the experience that Boethius goes through in his reflections on his misfortune that he learns to see as fortune. "The tongue that first has tasted bitter food/Finds honey that the bees have won more sweet." (Boethius 1983, 79) Boethius senses this with his own body as he describes it with this metaphor of taste, and in the lived experience of confinement. Philosophy offers him paradoxical knowledge to ponder and reflect upon. Good fortune seems to bring happiness but "deceives you with her smiles", whereas bad fortune is "always truthful because by changing she shows her

[4]Friedrich Nietzsche, *Menschliches Allzumenschliches* II, Vermischte Meinungen und Sprüche, § 170. *Kritische Studienausgabe der Werke Nietzsches*, Berlin, New York: de Gruyter, vol. 2, 449.

[5]In her Lectures on *Kant´s Political Philosophy* Hannah Arendt argues that Kant´s notion of aesthetic taste can be interpreted as a base for his understanding of taste as political judgment. (Arendt 1982).

true fickleness“ (Boethius 1983, 76). The type of philosophical thinking practiced here results in a judgment based on Boethius' new sense of his situation. Through thinking, Boethius comes also to understand his sensations and experience them in a new way. So the meaning of freeing himself from corporeal things cannot mean escaping into thought by dividing soul and body. It is rather through attunement to the emotions involved that Boethius frees himself from difficult passions and is able to gain a deeper meaning of fortune, and to see his misfortune in a consoling light. He displays a quality of mind that bridges between mind and spirit for both are embodied. In this sense Philosophy is a figure that brings together *logos* and heart, reason and emotion.

The Consolation of Philosophy is not only a literary dialogue but also most importantly a philosophical text. The word "text" is etymologically related to textile. Philosophy emphasizes that she herself weaves the textile for her robe. In her dialogue with Boethius they weave together different threads of thought, and so continue jointly weaving the philosophical robe by advancing philosophical knowledge and wisdom. These thoughts are not strictly cognitive philosophical assertions. They are affective thoughts as becomes apparent in the opening parts of the book when Philosophy dries the tears of Boethius with her dress, absorbs his tears with the cloth of her robe (Boethius 1983, 38). She says that he has for a while forgotten who he is, "but he will soon remember once he has recognized me" (Boethius 1983, 38). This statement may hold true for the whole of male-centric Western philosophy as it is represented in conventional canons and curricula of academic philosophy that omit women and the feminine. Philosophers have forgotten who they are since their origin in the wisdom of Philosophy has been erased or sublimated. The work of philosophy that needs to be undertaken is to remember our origins in Philosophy and all that she represents and can still represent once she has been recognized.

In a strange way, my reading of *The Consolation of Philosophy* also has to do with a belated recognition of the wisdom of my mother. When she gave me this book as I was embarking on the study of philosophy, an academic discipline she did not have much knowledge of, I think she had a hunch that this book that she had read may have a meaning and importance for me. This insight comes to me in a delayed fashion when I finally read and studied this book. My interpretation is not a specialist interpretation since both Boethius and early medieval philosophy are outside my areas of specialization within philosophy. Therefore, I have read *The Consolation of Philosophy* as a fresh encounter with philosophy, as if I was a beginner again, in an effort to recognize the Philosophia we as philosophers confess to love.

References

Arendt, H. (1978). *The life of the mind*. New York: Harcourt.

Arendt, H. (1982). In R. Beiner (Ed.), *Lectures on Kant's political philosophy*. Chicago: University of Chicago Press.

de Beauvoir, S. (2009). *The second sex*. London: Vintage.

Boethius, A. (1983). *The consolation of philosophy*. (V. E. Watts, Trans.). New York: Penguin.
Hehle, C. (2012). Boethius' influence on german literature to c.1500. In N. H. Kaylor, & P. E. Phillips (Eds.), *A companion to Boethius in the middle ages* (pp. 255–318). Leiden, Boston: Brill.
Irigaray, L. (1985). *Speculum of the other woman*. (G. C. Gill, Trans.). Ithaca: Cornell University Press.
Irigaray, L. (1993). Sorcerer love: A reading of Plato, *Symposium*, Diotima's Speech. In *An ethics of sexual difference*. (C. Burke, & G. Gill, Trans.) (pp. 20–33). Ithaca: Cornell University Press.
Jung, C. G. (2009). In S. Shamdasani (Ed.), *The red book. Liber novus*. New York: W.W. Norton & Co.
Kaylor, N. H., & Phillips, P. E. (Eds.). (2012). *A companion to Boethius in the middle ages*. Leiden, Boston: Brill.
Liddel, H. G., & Scott, R. (1996). *A greek-english Lexicon*. Oxford: Clarendon Press.
Nagel, T. (1989). *The view from nowhere*. Oxford: Oxford University Press.
Nash-Marshall, S. (2012). Boethius's influence on theology and metaphysics to the 1500 c. In N. H. Kaylor, & P. E. Phillips (Eds.), *A companion to Boethius in the middle ages* (pp. 163–192). Leiden, Boston: Brill.
Nye, A. (1989). The hidden host: Irigaray and diotima at plato's symposium. *Hypatia* (1989), 45–61.
de Pizan, C. (1999). *The book of the city of ladies*. (R. Brown-Grant, Trans.). London/New York: Penguin.
Schoeller, D., & Thorgeirsdottir, S. (2019). Embodied critical thinking: The experiential turn and its transformative aspects. *philoSophia* 9:1, 92–109.
Shanzer, D. (2009). Interpreting the consolation. In J. Marenbon (Ed.), *The Cambridge companion to Boethius* (pp. 228–254). Cambridge: Cambridge University Press.
Stopczyk-Pfundstein, A. (2003). *Sophias Leib. Der Körper als Quelle der Weisheit*. Stuttgart: BOD.
Thorgeirsdottir, S. (2020) Shame, vulnerability and philosophical thinking. *Sophia* , *59*(1), 5–17.
Tsakiridou, C. A. (1999). Philosophy abandons woman: Gender, orality and some literate pre-socratics. In E. Bianchi (Ed.), *Is feminist philosophy philosophy?* (pp. 234–263). Evanston: Northwestern University Press.
Waithe, M. E. (1987). Diotima of Mantinea. In M.E. Waithe, (Ed.), *A history of women philosophers* (Vol. 1). *Ancient women philosophers*. Dordrecht Netherlands: Martinus Nijhoff Publishers.

Chapter 7
A Journey of Transformative Living: A Female Daoist Reflection

Robin R. Wang

Health of body and tranquility of mind are the twin goals of philosophy's quest for a blessed life
Epicurus

Abstract By presenting the original works of female Daoist Cao Wenyi 曹文逸 (1039–1115) from the Song dynasty, this essay will examine how reality and human life were understood and what method was used by this female practitioner of Dao. Thus, this study might become a link for the cross-fertilization between the more abstract fields of philosophy and the concrete methodology of the study of women thinkers, leading to a comprehensive scheme for a better understanding of women's contribution to our knowledge in a global setting. In other words, the case of this female Daoist will invite us to a further methodological reflection: What does it mean to read those forgotten, marginalized women philosophers? How does it change philosophy? What do we learn from their thoughts and experiences?

In traditional Chinese philosophy pondering the ultimate questions about existence has always been intertwined with thinking through the body. Many early Chinese texts have illustrated that the natural world itself is conceived primarily as an organic human body. Daoism, a *Dao* based and nature-inspired teaching and practice, claims that the human body partakes in the shape and logic of the cosmos and the body can express the entirety of cosmic generations and transformations. Thus, human beings should act in ways that will bring about their alignment with the cosmos. This cosmic link partly justifies the kind of thinking that is related to the human body and the gender dynamics between man and woman. The *Huangdi Neijing* (黃帝內經) known as *The Yellow Emperor's Inner Classic* (111 CE) gives a naturalistic account of the human body. It rests on the following assumptions: the human body shares the

R. R. Wang (✉)
Department of Philosophy, Loyola Marymount University, One LMU Dr., 90045
Los Angeles, CA, USA
e-mail: Robin.Wang@lmu.edu

S. Thorgeirsdottir and R. E. Hagengruber (eds.), *Methodological Reflections on Women's Contribution and Influence in the History of Philosophy*, Women in the History of Philosophy and Sciences 3, https://doi.org/10.1007/978-3-030-44421-1_7

same categorical structure (*xianglei* 相類) with heaven, earth, and the myriad things; the pattern of human pathology is the same pattern as the changes of heaven and earth; human biological rhythm should change according to the rhythms of heaven, earth, and the four seasons. However, we could also say that the understanding of the natural world is modeled on the structure of the body, insofar as the cosmos itself is conceived on a biological model. These presuppositions also display a central belief in Chinese thought and culture: the unity of heaven, earth, and human being.

By presenting the original works of female Daoist Cao Wenyi 曹文逸 (1039–1119) from the Song dynasty, this paper will examine how reality and human life were understood and what method was used by this female practitioner of Dao. Thus, this study might become a link for the cross-fertilization between the more abstract fields of philosophy and the concrete methodology for a study of women thinkers, leading to a pervasive scheme for a better understanding of women's contribution to our knowledge in a global setting. In other words, the case of this female Daoist will invite us to a further methodological reflection: what does it mean to read those forgotten, marginalized women philosophers? How does it change philosophy? What do we learn from their thoughts and experiences?

Cao Wenyi whose Daoist name is Daochong, and nickname is Cao Xiangu, was seen as the 'master of tranquility and human virtue and the perfection of the Dao' (Despeaux 2003: p. 135). She has been a popular female figure through Daoist history and exalted by several Qing dynasty lineages of woman's inner alchemy. Cao was born into a wealthy family in Ningjing, Hebei. According to the record, she could read and cite poetry at the young age of five and was capable of remembering things after a mere glance at the age 15. By the age of 21, she defied the conventional path of arranged marriage and saw it as a form of imprisonment in this worldly life. She escaped an arranged marriage by fleeing to Yu Hua mountain where she led an extremely harsh and poor life. Despite these challenges, she contributed historical commentaries on the *Daodejing, Zhuangzi* and other Daoist texts. After gaining a well-known reputation as a Daoist practitioner, Cao was summoned to the capital of Kaifen, Hunan by Emperor Weizhong who was fond of Daoism and the literati. The emperor rewarded her with the title of 'Great Master of Literary Withdrawal into Clear Emptiness' (*Wenyi zhenren*).

Many of her writings have been lost, with the exception of a text called *The Song of Ultimate Source of Great Dao*, (*Lingyuan Dadaoge* 靈源大道歌) that has survived. This work is suitable for any beginner who wants to learn about Daoism and its practice. Cao's work was intended to inspire women and men to return to the great source of life, it 'acts as a guiding light to women throughout the ages' (Despeaux 2003: p. 140) to lead a transformative life. This work is the very first indisputable female Daoist work and hence the oldest known, authentic Daoist text written by a woman.[1]

[1] Although there is another text, *Huangting jing* 黃庭經 (*The Scripture of the Yellow Court*) which was credited to a woman Wèi Huácún 魏華存 (251–334), but it is debated if it is her own work. There are also female immortals such as Xie Ziran or He Xiangu in the Tang dynasty but none of them left any written texts.

The Song of Ultimate Source of Great Dao(*Dadaoge*)consists of 128 sentences written in poetic and elegant stanzas. Chen Yingning (1880–1969), the most influential twentieth century Daoist commentator and practitioner notes about her work in the context of Daoist writings, 'All Daoist texts from ancient to presents contain three things: (1) lavish discussions of mysterious efficacy 玄妙 (*xuanmiao*); (2) lively sayings 口訣 (*koujue*); (3) convincing and encouraging cultivation.' 修行 (*xiuxing*) (Chen 1988: p. 2). Following Chen's interpretation, we will discuss Cao's work from three aspects.

7.1 Lavish Discussions of Mysterious Efficacy (*bumiao*): The World as a Waterfall

This is how Cao starts the first sentence of *Dadaoge*:

> What I want to truly talk about is the very root of life.
> It comes from a genius source;
> Reflecting on one's longevity body is not empty;
> Spirited mirror (*lingjie*/body) contains myriad things;
> Ultimate (*taiji*) spreads mysterious? and human being gets the oneness;
> Received oneness should be carefully guarded and do not lose it. (Chen 1988: p. 5)
> 我爲諸君説端的, 命蒂從來在真息。
> 照體長生空不空, 靈鑿含天容萬物
> 太極布妙人得一, 得一善持謹勿失

This first passage reveals the Daoist worldview in the context of human body and one's mission in life. It relies on two traditional Chinese terms: Great Ultimate 太极 (*taiji*) and Mysterious Movement 布妙 (*bumiao*). The concept of the Great Ultimate gained popularity during Cao's time. The concept of *taiji* consists of three elements.[2] The first is that *tiaji* is the origin of the universe. Secondly, *taiji* contains two interrelated forces, namely *yin* and *yang* that have a generative efficacy on the myriad things.[3] The third element is that all myriad things are connected to the Oneness. The term *taiji* ultimately serves to explain how the world came into being with a conceptual coherence and consistency. The world is an ultimate oneness. We read from *Daodejing* 22 that the sage embraces oneness as the structure of the world.

[2]Lady Cao lived during the same period as Neo Confucian Zhou Dunyi 周敦颐 (1017–1073年) who wrote a short piece with a diagram called *Taiji Tushuo* (*The Great Ultimate Explained*). The term of *taiji* might be a popular concept for a broad understanding of the world. For more discussion see: Robin R. Wang, "Zhou Dunyi's *Diagram of the Supreme Ultimate Explained* (*Taijitu shuo*): A Construction of the Confucian Metaphysics", *Journal of History of Ideas*, December 2005, pp. 307–323.

[3]For a comprehensive study of yinyang see Robin R. Wang, *Yinyang: The Way of Heaven and Earth in Chinese Thought and Culture* (2012).

This vision and method of oneness is called guarding one 守一(*shouyi*), 'maintaining the one', or 'holding fast to the one' *(zhiyi)* (Moeller 2007).

The term for Mysterious Movement, *bumiao*, in Cao's work is also derived from 'The mysterious and the gate of all marvelous beings' found in the *Daodejing* Chap. 1 (Moeller 2007: p. 3). In fact, this mysterious gate refers to Dao, a generative living force as a female body.[4] *Bumiao* is the movement of *taiji*, mysterious and independent from the human will. Furthermore, *taiji* encompasses the self-generating, self-forming, and self-transforming forces leaving no need for an external cause or force to originate and sustain its existence. Cao tackled structural questions about the universe through concepts of *taiji* and *bumiao*, highlighting the oneness of myriad things and its self-generative force.

Cao's worldview continues a classical Daoist tradition similar to *Zhuangzi.* The world is like a waterfall that has infinite turbulences and uncertainty. The world in this view is chaotic,[5] random and embodied with probability. One can, however, work with these opportunities through her interactions with the different forces. Zhuangzi's dynamic view challenges the clock-like worldview. A precise, predictable and orderly clock cannot solve the issue of *sheng* 生, which is the generation of myriad things. How do the myriad things come into being? As Peter Corning puts it, 'Rules or laws, have no causal efficacy; they do not, in fact, generate anything' (Corning 2002: p. 19). Rules can merely describe regularities in phenomena and show consistent relationships in nature. What happens when multiple things interact with each other? Zhuangzi's waterfall is guided by probable reasoning that privileges one's own ability to assess situations and deal artfully with opportunities.

7.2 Lively Sayings (*koujue*): Knowing the World "Through the Handle of Life"

The waterfall worldview shifts one's focus towards the internal forces: *jing* 精 (essence), *qi* 氣 (energy) and *shen* 神 (spirit) and it places importance on self-empowerment and self-cultivation. As *Lu Shichunqiu* claims, 'The sages, having deeply contemplated the nature of the world, found nothing more worthy of esteem than life' (Knoblock and Reigel 2000: p. 80). Valuing life is the root of life, death, survival, and destruction. Therefore, the sage's own body is a storehouse of *jing* (essence), *qi* (energy) and *shen* (spirit). Her body is a resource and a reliable basis

---

[4] For more discussions on the connection between *Dao* and femininity see Wang, 'Dao Becomes Female: A Gendered Reality, Knowledge, and Strategy for Living', in: *The Routledge Companion to Feminist Philosophy*, eds. Ann Garry, Serene J. Khader, and Alison Stone, 2017.

[5] A similar conclusion about a chaotic world was reached in 1972. MIT's mathematical meteorologist Edward Lorenz made famous the 'butterfly effect' by asking 'Does the flap of a Butterfly's Wing's in Brazil Set off a Tornado in Texas?' It is a striking metaphor for chaos. (Gribbin 2004: p. 60).

for the sage's existence. *Zhuangzi*'s *zhenren* 真人 (a human of perfection)[6] is capable of lifting heaven and earth, grasping *yin* and *yang*, breathing true *qi* 氣, and depends on *shen* 神 (spirit) to enjoy longevity and the mastery of universal time.

These classical Daoist visions of human life were pursued and actualized in Cao's *Dadaoge*. Stanza 25–28 reads:

> *Shen* (spirit) is *xing* 性 (inner tendency), *qi* is *ming* 命 (lot)
> The spirit does not gallop away,
> *qi* will be firmly solidified.
> Originally they are intimated linked;
> If they ever are dispersed—what will be the primordial handle of life?
> Integrated to become one yet forget one,
> it can be changed with the primordial transformation
> 神是性兮氣是命, 神不外馳氣自定。
> 本來二物更誰親, 失卻將何爲本柄?
> 混合爲一複忘一, 可與元化同出沒。

Cao addresses two important notions in Daoist self-cultivation. Unlike Western distinctions between spiritual soul and material body or lofty rational mind and lowly emotional body, Cao perceives the human being as the unity of *xing* and *ming*. The concept of *xing* literally means 'inherent disposition' or 'innate tendency.' The word *xing*, however, appeared after the word for life 生 (*sheng*) so it has an intrinsic connection with *sheng*. *Xing* has two parts: *sheng* (life, generation) and *xin* 心 (heart/mind). While it refers to something fundamental, it is more like *psyche* than *soma*. *Xing* is also divided into two kinds: 先天 *xiantian* (prenatal heaven, as endowed by life itself) and 后天 *houtian* (postnatal heaven, as formed through one's lifetime).[7] This distinction is a key Daoist presupposition.

Another vital aspect of the human being is *ming*. *Ming* refers to one's vital force or circumstantial trajectories. It connects one's physical aspects such as one's biological conditions, life expectancy, or unexpected events that can occur. The notion of *ming* instantiates a recognition of our limitations in life. These limitations are imposed by a wider range of human conditions, such as one's physical capacities, health, temperament, emotional range, talents, society, culture and historical circumstances. To be wise is to be attentive to these limitations in the construction of one's life journey, as it would be both foolish to be willfully blind to them.

---

[6]In Chap. 6 Da Zhongsi, *Zhuangzi* provides a comprehensive description of *zhenren*. *Zhenren* is reaching a stage of true knowledge with extraordinary ability and emotions. Ultimately *zhenren* has unified with nature or cosmos (Ziporyn 2009).

[7]In the Confucian tradition, *xiantian xing* is given by heaven. Confucius states, '*Xing* is the same but its cultivation makes the difference.' In *Mengzi*, heaven's *xing* that animates every human being is manifest in four sprouts in human *xin* 心 (heart/mind). Human beings thus are advised to 'preserve the *xin* to cultivate the *xing*.' The *Zhongyong* (one of the Confucian classics) claims that 'leading by the *xing* is called *dao*.' 率性之謂道. The *Yijing* states,'One *yin* and one *yang* is called *dao*. Led by it is goodness (*shan*); completed by it is the *xing*' (成之謂性). In this sense the *xing* is the result of *yin* and *yang* interaction. In Buddhist teaching *xing* is also something with paramount importance. Those who are able to grasp the *xing* will become Buddha. (*Jianxing chen fu* 見性成佛)

Ming is given by heaven[8] and connected with one's *xing* 形 (form) or physical body, while *xing* is related to the one's *xin* 心 (heart/mind). The connection between *xing* and *ming* has become the testing ground for different Daoist schools. In early Chinese texts, the compound *xingming* 性命 is defined as 'the two overlapping factors that together determine life's course.' They are two aspects of the same human life or the same principle. So the cultivation of life refers to conjoined cultivation of *xing* and *ming*.

In *Zhuangzi*'s story of the waterfall (Ziporyn 2009: p. 81) the Man was born on the hill, grew up in the water and was able to follow his *ming*. Zhuangzi grafts a Daoist view of *xing* and *ming* from the waterfall story: 'Being in the water and having peace in the water is *xing*; knowing and not knowing the reason is *ming*' (Ibid.). His observation suggests that Daoist *xing* represents tranquility in the human mind, which can be cultivated. *Ming* on the other hand, represents the unknown facts in human life that nevertheless should be embraced.[9]

To live a flourishing life one must be able to know the *ming* and cultivate the *xing* 知命修性 (*zhiming xiuxing*). These are shared presuppositions common to Confucians, Buddhists and Daoists. For each, there are two different methods to fulfill the two aspects of the human condition.[10] But Daoists took a different path than Buddhists and Confucians by emphasizing the synchronized cultivation of *xing* and *ming*. More importantly, Daoists have described their theory and practices as teaching of *Xingming* 性命學 (*xing ming xue*).

Within this Daoist tradition Cao focuses on the dual cultivation of *xing* and *ming*, where one should cultivate *xing* and *ming* at the same time. Cao makes a significant distinction by identifying *shen* with *xing* and *qi* with *ming*. She explains *xing* as *shen* (spirit) and *ming* as *qi*, (energy flow or life force). These dual processes bring one's mind, body and spirit into a stage of 'clarity and tranquility' 清靜 (*qingjing*). This unification has been recognized as Cao's valuable and unique contribution to the Daoist history of teachings of *Xingming*.

---

[8]Confucius thinks that 'life and death depend on the *ming*.' He also perceives the importance of knowing the *ming*. It took him 50 years to know the *ming* of heaven (50 知天命) *zhi tianming*, *Analects* chapter two.

[9]*Xing* and *ming* roughly as *psyche* and *soma* are two significant constituents of the human being. According to the *Shangshu* 尚書 (The *Book of History*) they have a direct effect on the five *fu* 五富 (fortune/happiness) and the six worries 六機 (*liuji*) in human life. The five fortunes are longevity 壽 (*shou*), wealth 富(fu), peace 康寧(*kangning*), love of virtue 好德 (*haode*) and long fortune 考終命 (*kao zhongming*). The six worries are ominous 凶 (*xiong*), short 短 (*duan*), broken 折 (*zhe*), sickness 疾 (*ji*), depression 憂 (*you*), poverty 貧 (*ping*), evil 惡 (*e*) and weakness 弱 (*ruo*). Those who enjoy the heavenly *xing* and know *ming* will be blessed with the five types of happiness and will steer clear of these six worries.

[10]For Confucians this is a social project. Human realization and knowing the *xingming* depend on the ritual practice. This ritual practice is to achieve a social and communal harmony (*Analects*, 1.12). A person's seven 情*qing* emotional states, *xi* 喜 (happy), *nu* 怒 (angry), *ai* 哀 (sad), *lu* (pleasure), *ai* 愛 (love), *e*惡 (evil) and *yu* 欲(desire) are embedded in his or her *xing*. Thus one should cultivate a proper relation between *qing* and *xing* through ritual. The ten *yi* (righteousness) are also contained in *ming*. Confucians, accordingly, promote a human being's social and mental cultivation. For Buddhism *ming* should be eliminated through cultivation of *xing*.

The unity of *shen* and *qi* or *xing* and *ming* as the 'handle of life' demonstrates the critical role of body cultivation in Daoism. The 'handle of life' can only develop through the cultivation of the body-mind. In *Dadaoge*, Cao takes human mind as numinous mirror 靈鑒(*lingjie*), (*Dadaoge*, 4) which contains and reflects all myriad things,[11] and human body as numinous residence 靈府 (*lingfu*), (*Dadaoge*, 8) where it is full of chambers and palaces for universal spirit and various gods to reside.

According to Cao, one of the first conditions for the embodied human being is that mind, the numinous mirror, must be emptied through cultivation for spirit (*shen*) to remain. Cao goes on:

> Focusing *qi* to its ultimate softness *shen* will stay forever;
>
> The true *qi* will come and go freely.
>
> All *qi* will constantly return to the primordial source of life (*yuanming*). (Dadaoge, 41–43)
>
> 專氣致柔神久留, 往來真息自悠悠。
>
> 綿綿迤邐歸元命.

The second condition is that the body, which is the numinous residence or lodging place, must be clean and proper to allow the spirits to come and stay. This function for achieving a numinous body relies on self-cultivation:

> One should cultivate *shen* and *qi*.
>
> It will be a wasted effort if *shen* and *qi* are not peacefully settled.
>
> It is a pity to have a wonderful place, just as a golden palace without the master residing in it.
>
> One should use different methods to call back the master (spirit) Let the master stay without the practical functions (Dadaoge, 117–122).
>
> 比來修煉ɹ神氣, 神氣不安空苦辛。
>
> 可憐一個好基址, 金殿玉堂無主人。
>
> 勸得主人長久住, 置在虛閑無用處。

The cultivation of the body is called *xiuzhen* 修真 (cultivation of perfection) in order to become *zhenren*, (a human of perfection). Cao asks:

> Where is the *zhenren* coming from?
>
> *Zhenren* is originated within our own numinous place;
>
> In the past it was deeply covered by clouds,
>
> Today it is seen through the eye of Dao.
>
> This is not the work of one day or one night,
>
> This is not the skill but a pure self. (*zhenwo*) (Dadaoge, 51–56)
>
> 借問真人何處來? 從前原只在靈台。
>
> 昔年雲霧深遮蔽, 今日相逢道眼開。
>
> 此非一朝與一夕, 是我本真不是術。

---

[11] Mind as mirror metaphor is a common trend in Daoist teaching since Zhuangzi's 'fasting of the mind' theory (Ziporyn 2009: p. 26).

Here again, the human being contains the seed of goodness or the seed of *zhenren*. However, this goodness must be activated through the effort one exerts over a lifetime. To open the eyes of Dao is to see the true self being buried in clouds. These methods can only be done properly through Daoist practices. It is believed that there is a genuine and valuable germ in the human body that is something divine, spiritual, transformative and important. As such, one must learn how to take care of it from moments of conceptual understandings of the Dao to developing a physical regiment. Daoism provides instruction on how to attain those moments of understanding along with physical measures for this transformation.

For Cao, physical cultivation is an important aspect of living. Knowing and doing are equally essential components to become *zhenren*. The body contains constructive cosmic, spiritual and epistemological dimensions. A wide range of human concerns such as emotions, social problems, and death can all be explained without appealing to an external object, such as a transcendental being. The anxiety of living different cycles and facing the troubles of life can be quieted by focusing on one's self-cultivation.

> If the place is empty then *shen* will stay and relax.
>
> *Lingfu* (numinous residence) is fried and exhausted and will dry up the blood flow;
>
> Sadness, pleasures, worries, unwinding and winding all damage and rot the body.
>
> Day in and day out one does not know where to go,
>
> The spirit is lost and has no foundation.
>
> One gradually becomes exhausted,
>
> In the end, primordial *qi* is worn out and only *shen* is left. (Dadaoge, 8–14)
>
> 宮室虛閑神自居, 靈府煎熬枯血液。
>
> 一悲一喜一思慮, 一縱一勞形蠹弊。
>
> 朝傷暮損迷不知, 喪亂精神無所據。
>
> 細細消磨漸漸衰, 耗竭元和神乃去。

It is a common assumption in the Daoist teachings that health can be degenerated and the spirit damaged by emotions. The way to avoid physical degeneration and a damaged spirit is to be with Dao by cultivating the mind and body. As such, the living and sentient body is celebrated as the organizing core of intellectual, spiritual and practical experience. The body is not only a crucial space where one's ethos and values are displayed but also where one's cognitive capacities can be enhanced. Cao says:

> Expanding and melting vessels to transform nerves and bones;
>
> Brightness is reflecting everywhere with a complete penetration.
>
> Three bugs be expelled out of the body,
>
> So ten thousand spirits will come into the red palace (Dadaoge, 47–50).
>
> 蒸融關脈變筋骨, 處處光明無不通。
>
> 三彭走出陰屍宅, 萬國來朝赤帝宮.

In the Daoist teaching, three harmful bugs (*sanchong*) or ghosts can reside in three central human body parts. The upper bug is in the head. It blocks one's wisdom and makes one intellectually inept or stupid. The middle bug is located in one's chest. It compiles emotional disturbances and entanglements and does not allow for peace and calm. The lowest bug hinders one's stomach. It causes the desire for food and a lust for sex.

When a perfect natural body has been eaten away by some of these three dangerous bugs one must follow Daoist teaching and practices to get rid of them. If the body goes through a transformation, then these bugs have no place to reside any longer. Once the three harmful bugs have been removed, then healthy *jing*, *qi* and *shen* are restored. One of the imperative reasons to cultivate the body is that ten thousand spirits can inhabit the cleansed body.

This Daoist practice requires special mental cultivation and enlightening. For Cao, thinking is an impediment to attain a well-ordered mind. One does not simply enter a quiet physical space to become centered. In solitude, one must find delight and brightness in one's own body by enlarging the mind, letting go and relaxing the *qi* for expansion. In this state, the body is calm and unmoving. Oneness guides the body for it to discard the myriad things, thoughts and wills, which disturb it. One does not see, nor is enticed, by profit. Ultimately one aims to arrive at a fittingness of her body, mind, and spirit to the Dao. Evidently, Cao pays more attention to ways for inner cultivation. She suggests that human energy relies on the refinement of one's own body/mind.

When one gets the Dao, one will be able to penetrate the patterns of *li*; then one will possess a clear-head for *quan*, that is, the weighing of things. It is a way of life organized by physical and mental hygienic principles. There is a systematic process of negating, forgetting or emptying out the contents of consciousness. Clearing consciousness includes elements such as perception, emotions, desires, thoughts and even linguistic distinctions.

7.3 Convincing and Encouraging Cultivation: Life as the Journey Between Transcending and Returning

Even though there are many differences between the Daoist tradition and other philosophical traditions, one shared concern is the meaning of human life. *A life of grateful striving* is a shared vision and goal that serves as a common ground between divergent traditions. What makes human life meaningful? Must the meaning of life be derived from the notion of transcendence?

The absence of transcendence in Daoist thought does not drive it into either pure naturalism or materialism. A strong commitment to the theory and practice of transcending current human conditions is a permanent concern in Cao's work, indicated by her saying:

> Knowing a great Dao is not easy to put into practice.
>
> Fame, achievement and body are hard to escape.
>
> Finding a quiet place to practice,

> Getting all to cultivate the great focus.
> Physical form and spirit have to be complete,
> If physical form cannot be first completed one must complete spiritual form.
> Do not seek fame or compete profits,
> Ending worldly human emotions to attend no-events stage (Dadaoge, 95–122).
> 心知大道不能行, 名迹與身爲大病。
> 比如閑處用功夫, 爭是泰然修大定。
> 形神雖曰兩難合, 了命未能先了性。
> 不去奔名與逐利, 絕了人情總無事。

Cao calls for practices that move beyond mundane existence toward a greater spiritual realm, reaching out for the *Dao*. It intends to prepare the body for higher stages. The Daoist attainment is for "an energetic transfiguration that leads to ultimate transcendence and inner spiritualization" (Kohn 2008: p. 31).

Cao's method is a great way for transcending the limited human condition. It is a process of going beyond. Perhaps it can be seen as transcending 超越 (*chaoyue*). Indeed this process differs from transcendence as a metaphysical conception or framework. Transcending in Daoist practice is an embodied practical development and a process of transformation. It has a strong conceptual commitment where dedicated practice enables one to go beyond the current human physical, mental and social conditions. It is a course of action that is not a state of being, but a journey over the span of one's life.

The Daoist's journey is composed of a few designated goals. First, the ways in which one can make the body a lodging place for spirits or natural goodness. In order to make this possible one's heart/mind (*xin*) has to be in fit condition and *qi* sufficient. This life-giving *qi* comes from nature or cosmos. One needs to maintain communication and a deep relationship with nature's *qi*. This communication or association is the foundation of life.

The second goal is to realize that this journey is a journey of returning. This particular goal is to demonstrate where the body comes from and where the body is going. *Taiping jing* (*Classics of Great Peace*) claims:

> At the moment when one gets a life one is separated with heaven and earth (*feng sheng*) and embraces *yuanqi* to be a human being. Before that time [one has not divided with heaven and earth], one does not need to eat or drink it, just takes *yinyang qi* to survive that does not even know the hungry and thirsty. But then one gradually leaves the way of spirit farther and loses the meaning of spirit. One does not know how to return to Dao's emptiness but is always away from it. (Yang 2002: p. 106).

This is a picture of human life and why there is a need to return to the root of Dao. In sum, the stages can be seen as first, getting a form of life that is separated from the great one. At the beginning of this life, one does not depend on external materials such as food and water. The life cycle can then move away from the Dao or undergo the regenerative process. For it, one needs to cultivate his/her body and let a divine fetus form and flourish within it. The ultimate goal is to return to nature and be one with the Dao, going from a state of separated body 分身 (*fen shen*) to a untied body 合身 (*he shen*).

The third goal is an enjoyable and imaginative journey. It is neither restraining nor restricting bodily discipline. By the end one can penetrate gold and pass through rocks. This means that one is no longer limited or restricted by bodily forms and spirit wanders freely. In the end, one gains a pervading spirit 神通 (*shentong*) that is materialized human imagination. It is a supernatural power. Daoist practice raises awareness of inner energy flowing outside patterns of vibration. Beginning with the manipulation of bodily energies, it leads to the ultimate spiritual achievement. Physical cultivation is a life long journey that entails persistence and constant diligence. Yet it is a rewarding voyage. Cao ends the *Dadaoge* with confidence:

> A Persistent and unified heart/mind will walk towards future,
> Eventually, the great Dao will not disappoint human beings. (Dadaoge, 127–128)
> 堅心一志任前程, 大道於人終不負

7.4 Conclusion

This paper endeavors to present a female Daoist work which is hardly known in the field of Daoist study both in China as well as in West today. If philosophy aims to pursue wisdom then it seems only natural and logical to embrace a diversified "voice of reason," including those overlooked or disregarded voices from women. Cao's work makes a valuable stride in enlarging the perspectives of contemporary philosophical discourse. Cao affirms a remarkable insight into the connection between mind and body although her work was not intended to change women's social and political position in China. Her work challenges the mind-body asymmetry where the mind poses as a disembodied universality while the body is constructed as disavowed corporeality. The mind is not based on the exclusion of body nor does the body entail a rejection of mind. There is no body outside of the mind and there is no mind outside of the body. This prescribes a developmental and dynamic process that defines an original fullness of the ultimate reality and the human being. There is a feminine wisdom tradition in her writings, an emphasis on the body, on relations to others, which yields a richer idea of the human being on the basis of accounting for the sexual differences of men and women. Her metaphor of pregnancy is also remarkable insofar as she describes being one with Dao as having a divine fetus form and flourish within. With this metaphor, she alludes to how being one (with Dao) includes being two within as having some other being nurtured within. Otherness is hence an integral part of the self. The subject is an other to itself when it is at one with Dao. This idea of a relational self reminds us of contemporary feminist notions of the self, such as Luce Irigaray's conception of woman as the sex which is not one (Irigaray 1977: pp. 23–33).

Cao's work also reminds us again of the interconnectedness of human existence and the complexity of a multilayered human understanding. It needs a vital imperative for re-appropriating canonical texts, re-examining historical and religious contexts and re-imaging the conceptual and practical resources. With the rise of Asian/Chinese philosophy in western philosophy departments and professional associations, women philosophers of the Chinese tradition should be read not left out. Their works should be rediscovered and taken into the canon and curriculum of Chinese philosophy. In fact, female Daoists are being quietly revived in China today. One example is the *Academy of Female Daoists* at the Southern Mountain (*Nanyue* 南岳) in Hunan, China. Nanyue is one of five central Daoist mountains, with a long history of living masters. Today at the bottom of this mountain, there is a two-year on-site academy committed to train women Daoists. This academy and Nanyue are both under the leadership of a well-known female master Huang Zhian 黄至安, president of the Hunan Daoist Association, who also serves as the vice-president of the Chinese Daoist Association. She is one of the most prominent female Daoists in China today, at least in terms of political and social involvement. She is also the chief editor of the biannual journal *Hunan Daoism* 湖南道, which contains ten sections that range from discussions of current events through presentations of classical texts and Daoist history to contemporary moral issues. Master Huang has a long-term vision and firm commitment to promote Daoism in the contemporary world. With all kinds of funding, she has acquired a large abandoned elementary school building and turned it to the site of a women's academy. Its mission is summarized in the following slogan: *Honoring Dao, Respecting De* (Virtue); *Learning and Cultivation Progressing Together* 尊道贵德, 学修并进. The goal of the program is to train Daoist leaders who can manage and run temples all over the country and participate in a wide range of social activities. This newly trained elite is set to become a vigorous force affecting the development of Daoism in the coming decades.

From classical Daoist texts like Cao's to contemporary female Daoist practices we can recognize a more emancipatory and enriching potential source for women than in mainstream/malestream Daoism. More importantly, embodied thinking, like Cao's work represents, is a core feature of Daoist philosophy. Cao's writings can be seen as a kind of female consciousness of *Dao*, or a Daoist female consciousness. Her work expands, supports, or alters our assumptions about the way of thinking or philosophy. The overarching focal point of this understanding lies in a special way of taking philosophy not only as a rational argumentation or practical social-political thought but also as a deep engagement with one's body and embodied transformation. Cao Wenyi has described the ways for us to access that wisdom and to avoid the 'bugs' in the head to be stupid and in the stomach to be feared.

References

Chen, Y. (1988). *Interpretations and commentary on the song of great Dao of Lingyuang, Daoist Association Publishing or Commentary on Lingyuan Dadaoge*. Beijing: Chinese Daoist Association Publishing.

Corning, P. A. (2002). The re-emergence of emergence: A venerable concept in search of a theory. *Complexity, 7*(6), 18–30.

Despeaux, C., & Kohn, L. (2003). *Women in Daoism*. Cambridge: Three Pines Press.

Gribbin, J. (2004). *Deep simplicity: Bring order to chaos and complexity*. New York: Random House.

Irigaray, L. (1977). *This sex which is not one*. Ithaca: Cornell University Press.

Kohn, L. (2008). *Daoist body cultivation*. New Mexico: Three Pines Press.

Knoblock, J., & Reigel, J. (2000). *The annals of Lü Buwei: A complete translations and study*. Stanford: Stanford University Press.

Moeller, H.-G. (2007). *Daodejing: A complete translation and commentary*. Chicago and La Salle: Open Court.

Wang, R. R. (2012). *Yinyang the way of heaven and earth in Chinese thought and culture*. Cambridge: Cambridge University Press.

Yang, Jilin 杨寄林 (Ed.). (2002). *Taipingjing* 太平經 *Classic of great peace*. Shijiazhuan: Hebei People's Press 河北人民出版社.

Ziporyn, B. (2009). *Zhuangzi: The essential writings with selections from traditional commentaries*. Indianapolis/Cambridge: Hackett Publishing Company.

Part III
Reflecting the Content

Chapter 8
Reconsidering Beauvoir's Hegelianism

Karen Green

Abstract This paper argues that the widespread Hegelian legacy that feminism has inherited from Beauvoir is highly problematic and that feminists, in particular, should be suspicious of philosophies of history and histories of philosophy that take Hegel too seriously. Any such history or philosophy will fail to take into account the deep roots of women's comparatively equal status in the West in the long history of women's political, ethical, theological, and philosophical theorizing since the fifteenth century. Nevertheless, in a reformulation of Beauvoir's Hegelianism, this paper proposes a new historical dialectic understood as a conversation in which female and male voices are equally represented and the unfolding of spirit is transformed into a dialectic of sex.

The observation that, with regard to their political status, married women are apparently no better than slaves, harkens back to before Mary Astell's rhetorical, "If all men are born free, how is it that all women are born slaves?" (Astell 1706, Preface). In *The Second Sex*, Beauvoir implicitly extends this metaphor, by borrowing Hegel's analysis of slave consciousness, in order to argue that women experience themselves as objects, as other, in a manner analogous to the Hegelian slave. The idea of the oppressed as the other, and the language of objectification have become commonplaces of radical philosophy in one form or another. Such forms range from the writings of Beauvoir's somewhat critical feminist daughters, to the wide dissemination of Sartre's existentialist application, to the case of anti-semitism and racism, of the Hegelian account of oppressed consciousness. In this paper I argue that this widespread Hegelian legacy is highly problematic and that feminists, in particular, should be suspicious of philosophies of history, and histories of philosophy that take Hegel seriously. Any such histories or philosophies fail to take into account

K. Green (✉)
SHAPS, University of Melbourne, Parkville, Melbourne 3010, Australia

School of Historical and Philosophical Studies, University of Melbourne, Melbourne, Australia
e-mail: karen.green@unimelb.edu.au

S. Thorgeirsdottir and R. E. Hagengruber (eds.), *Methodological Reflections on Women's Contribution and Influence in the History of Philosophy*, Women in the History of Philosophy and Sciences 3, https://doi.org/10.1007/978-3-030-44421-1_8

the deep roots of women's comparatively equal status in the West, in the long history of women's political, ethical, theological, and philosophical theorizing since the fifteenth century.

Simone de Beauvoir's *The Second Sex* deserves to be recognized as the fundamental source for late-twentieth-century academic feminism, which emerged in the 1970s. This claim might be contested, for by the 1970s, many feminist writers were already representing themselves as overthrowing, or going beyond Beauvoir. Luce Irigaray, for instance, came to represent Beauvoir as a feminist of equality, determined to eradicate sexual difference, which Irigaray defended (Irigaray 1991). Nevertheless, in *Speculum of the Other Woman* and *The Sex which is Not One* she retained and developed in a radical direction, Beauvoir's claim that woman is the other of man, an object to his subject (Irigaray 1985a, b). Irigaray accepts Beauvoir's 'woman is other' but rejects the proposal that women should strive for transcendence as characterized by Beauvoir, which she associates with masculine subjectivity. In a similar vein, Genevieve Lloyd, in *The Man of Reason* criticized Beauvoir's aspiration that women should achieve transcendence, suggesting that, as it occurs in Hegel it is, "… in its origins ….a transcendence *of* the feminine" (Lloyd 1984: p. 101). Nevertheless, the Pythagorean table of opposites, which Lloyd cites as evidence for the claim that 'maleness was aligned with active, determinate form, femaleness with passive indeterminate matter' (Lloyd, p. 194, n.3) had earlier been cited by Beauvoir, in her account of the mythical construction of woman as the other of man (Beauvoir 1949: pp. 1:313, 2:137). Both Irigaray and Lloyd thus build on Beauvoir's account of women's situation, while criticizing what they take to be her positive recommendations. Judith Butler, by contrast, took up the active, transforming, and emancipatory aspects of Beauvoir's claim that 'one is not born but rather becomes a woman' suggesting that it provides 'a potentially radical understanding of gender' in which the body becomes a 'field of cultural possibilities' (Butler 1986: pp. 35, 49; 1990).

Additional examples could be offered, but I hope that these three will suffice to establish that Beauvoir was the fundamental source for quite diverse strands of late-twentieth-century academic feminism and gender studies. Now, it is also clear that Hegel plays an ineliminable part in Beauvoir's account of the situation of women. During the first year of the Second World War, alone in Paris, while Sartre was a prisoner of war, Beauvoir had time to study Hegel's *Phenomenology of Spirit* in detail, and in her letters, she encouraged Sartre to confront his concepts of nothingness, being-in-itself, and being-for-itself with Hegel's more positive account of the unfolding of consciousness. This advocacy resulted in the chapters on concrete relations with others, found in *Being and Nothingness* (Green and Roffey 2010; Beauvoir 1990: pp. 81–82; Sartre 1943, 1983). The engagement with Hegel's *Phenomenology of Spirit* is, in turn, fundamental for Beauvoir's account of women's situation. Wondering why women have accepted their inferior position, she says that things become clear 'if, following Hegel, we find in consciousness itself a fundamental hostility towards every other consciousness; the subject only posits itself by opposing itself to something: it aspires to set itself up as essential and to constitute the other as inessential, as object' (Beauvoir 1949: p. 1:18).[1] And later, she offers the key to the mystery. The species maintains itself biologically by recreating itself, but this is merely the repetition of differing forms of the same life. It is in transcending life by means of existence, that man assures the repetition of life, and through this surpassing he creates values, which deny the value of pure repetition (Beauvoir 1949: pp. 1:115–16). So,

> Certain passages of the dialectic in which Hegel defines the relationship of the master to the slave would apply far better to the relationship of man to woman. The advantage of the master, he says, comes from his affirmation of Spirit as against Life through the fact that he risks his life: but in fact the conquered slave has known the same risk. Whereas woman is fundamentally an existent who gives Life and does not risk *her* life; there has never been a combat between her and the male (Beauvoir 1949, 1:116).

So, the whole key to the mystery of women's historical subordination rests on the fact that, because they have been 'biologically destined for the repetition of life' women's consciousness has become a 'dependent consciousness for whom the essential reality is the animal life: that is to say, a being given by another entity'. A reading of Hegel's dialectic of master and slave is essential for Beauvoir's thought, and this, in its turn, is the foundation of much late-twentieth-century feminist philosophy. Yet this dependence on Hegel's historical dialectic is, I hope to show, highly problematic for feminists.

The problem is, to put matters simply, that the account of history, according to which man affirms spirit, while woman merely repeats life, is itself a repetition of self-reproducing philosophical mythology. Hegel takes the history of ideas, the history of the evolution of the state, and the history of the progress of spirit, to be the history of men's ideas, the history of male political institutions, the history of men's consciousness. Men, it seems, have a need to represent themselves as, in some sense, superior to women. Whether this is an ideology that has developed in order to justify a legacy of oppression, grounded in historical differential access to physical force, or whether it is rooted in a deep-seated psychological need, I shall not attempt to determine. But men's accounts of their natural, cultural, and spiritual superiority to women are ubiquitous.

The myth of the male birth of culture gets one of its earliest philosophical articulations in Plato's *Symposium*, in which Diotima represents love as the desire to repeat and eternally bring forth beauty, and contrasts the animal repetition of mere biological reproduction to the aspirations of higher souls who wish to create cities and laws. A similar idea, that the male contributes a 'higher' active, spiritual element to formless matter, was also expressed in Aristotle's theory of generation, according to which, in biological reproduction, while male and female are members of the same species, the male contributes the active heat of generation, while the female contributes the passive matter, so that the female is formed as a result of a failure of the male semen to master the female menses (Aristotle 1941: pp. 675–676; Deslauriers 1998: p. 159). As a result of this theory, Medieval interpreters of Aristotle, who

[1] I am using my own translations of the first edition of *Le deuxième sexe*, in this and following passages from Beauvoir.

fused his theories with Galenic accounts of the four humors, represented femaleness as a defect, a lack, caused by the absence of sufficient heat and nourishment necessary for the full expression of the human form. At least as early as Christine de Pizan, the implausibility of this myth was pointed out by a woman. Christine followed Aristotle, in insisting that men and women are members of the same species, and argued that since God, and his servant nature, do not systematically make evil and deformed things, sexual difference is clearly natural and intended (Pizan 1997, 2001: pp. 13–14; Paakinen 2016). So there must be equally good forms of the male and the female body. Sexual difference is not a defect or kind of inferiority. By distinguishing the soul from the form of the body, she is able to represent both men and women as equally good, but different kinds of human creature, into whose bodies, God, or his servant nature, inserts immaterial souls, each of which is the image of God. Christine also postulated that it was her mother's strong desire for a girl child, which caused her body to take a female form, implying, against Aristotle, an active female principle of generation, and moving closer to the modern account, according to which biological sex is the outcome of competition between genetic elements, provided equally by the mother and the father (Pizan 1959). No one any longer takes the Medieval interpretation of Aristotle's theory of generation seriously, we now understand that, in generation, both the male and the female contribute half the DNA that is twisted together in a double helix. Although, like Aristotle, we do accept that the sex of the child is determined by the semen, like Christine we reject the idea that sexual difference implies inferiority. Yet, it seems we are still fascinated by an account of cultural reproduction, according to which it is only men whose words and writing constitute our intellectual history and it is men who have determined the shape or form of culture, while women make up the matter that has been shaped and formed. This is true of Beauvoir, as well as the tradition of the history of ideas into which she was initiated.

As I argued more than 20 years ago, by rearticulating the concept of woman as slave, through an adapted Hegelianism, in which women's consciousness is 'a dependent consciousness for whom the essential reality is the animal life,' Beauvoir tended to discourage research into the history of those women who had attempted to achieve transcendence (Green 1995: pp. 130–136). She claimed that women have 'never opposed female values to male values' and so, she is left to explore the myths through which men have defined women (Beauvoir 1949: pp. 1:117). This theory of oppression, according to which the oppressed only have access to an identity that rests on the objectification of the oppressor, was partly inspired by Richard Wright and also applied by Sartre to the situation of the Jews (Simons 1999; Sartre 1954). It has the awkward consequence of making the articulation of an authentic positive identity by the oppressed seem impossible, for it puts them in a double bind and faces them with a choice between equally unpalatable identities, as either objectified or objectifier (Green 1999a, b).

In her account of the Middle Ages, Beauvoir acknowledges that Christine de Pizan 'took up her pen to defend her sex' (Beavoir 1949: p. 1:177). Yet she concludes the historical section of *The Second Sex* with the overarching assertion,

> ... all of the history of women has been made by men. . . .men have created values, manners, religions; women have never disputed this empire with them. Some isolated individuals—Sapho, Christine de Pisan, Mary Wollstonecraft, Olympe de Gouges—have protested against the harshness of their destiny; and sometimes collective demonstrations have emerged: but the Roman matrons who joined together against the Oppian Law or the Anglo-Saxon suffragettes would never have been able to apply such pressure but for the fact that men were disposed to submit to it (p. 1:222).

But this capitulation to a Hegelian philosophy of history, according to which the history of philosophy is the history of a masculine spirit, succumbs to a theory of the reproduction of culture, that I believe is as distorted at the cultural level as is the Aristotelian theory of biological generation, at the biological level.

Ultimately, and particularly through their role as the first interlocutors and educators of children, the cultural DNA of every society must reproduce itself through its articulation by women as well as by men (Green 1989: pp. 85–96). Anthropologists have come to recognize this, and have begun to include women's representations of their culture, and their roles within it, alongside those of men (Bell 1983; Sanday 1981). But when one is dealing with sophisticated written cultures, where cultural reproduction has taken place through the repetition of texts, and women have been largely excluded from the institutions devoted to such transmission and repetition, it becomes difficult to retrieve the female contribution to cultural evolution. This has resulted in the intractable, fundamentally Hegelian history of ideas, which still dominates. One needs only to look at work such as Knud Haakonssen's *The Cambridge History of Eighteenth-Century Philosophy* to see how persistent it is (Haakonssen 2006). This work concludes with a bibliographical appendix that lists the names and works of 193 eighteenth-century philosophers, of whom just two are women: Émilie Du Châtelet and Mary Wollstonecraft. Our common knowledge follows a similar male trajectory. Moving from the Middle Ages to the end of the Enlightenment, we know more or less what the list of male philosophers looks like. The following does not claim to be absolutely authoritative but is representative: Abelard, Aquinas, Erasmus, Montaigne, Descartes, Malebranche, Cudworth, Spinoza, Leibniz, Newton, Locke, Hume, Kant.

Yet, although it is still not sufficiently integrated into the mainstream of philosophy, an enormous amount of work has been achieved over the past twenty or thirty years, which has added to this malestream history the achievements of a number of women philosophers. They are often inserted into the history of ideas through their relationships with men. Female philosophers become danglers on a chain of historical development that is nevertheless still represented as a dialectic of male thinkers. Héloïse, of course, has always been known as the famous consort of Abelard; Christine de Pizan translated and used some texts of Aquinas in her *Advision Cristine*; Marguerite de Navarre's defense of marriage might be usefully compared to that of Erasmus; Marie de Gournay gets remembered as the editor and 'adoptive daughter' of Montaigne; Elizabeth of Bohemia famously corresponded with Descartes; Astell was influenced by Malebranche; Damaris Masham was Ralph Cudworth's daughter; George Eliot borrowed from Spinoza; Emily Du Châtelet translated Newton and adapted Leibniz; Catharine Cockburn defended Locke; Catharine Macaulay attacked

Hume. To be added as minor jewels, glittering along the chain of masculine links is already an advance. But, once one begins to do the work, other possibilities begin to emerge.

I have to admit that, 20 years ago, I was profoundly ignorant of women's intellectual history, even of that within the culture to which I belong. My first book, *The Woman of Reason* jumped from Christine de Pizan to Mary Wollstonecraft, leaving between them a great gap of nearly 400 years. But since then, work undertaken with Jacqueline Broad and other collaborators on the history of women's political thought has begun to reveal an alternative dialectic of women's ideas in which later women recognize, respond to, criticize, and carry on the intellectual legacy of their foremothers (Broad and Green 2009; Green 2014). Beginning with Christine de Pizan, there is a line of filiation that takes us from her defense of women to the problem of the relative virtues and vices of the sexes, found in Marguerite de Navarre. Pizan's attack on Jean de Meung in the debate over the *Romance of the Rose* is taken up in the *Heptameron,* Marguerite de Navarre's suite of stories in which she allows her imaginary participants to debate the relative virtues and vices of men and women (Hicks 1977; Navarre 1999). It has not been possible to demonstrate that Christine de Pizan's works defending women were available in Italy, but the rich vein of Renaissance women writers there—Nogarola (2004), Cereta (1997), Fonte (1997), and Marinella (1999), the last of whom argued for the moral superiority of women—were known to Marie de Gournay. Gournay, famously, was content to claim equality with men (Gournay 2002a, b). Gournay and Marguerite de Navarre were clearly known to Madeleine de Scudéry, who recognized the literary contributions of Marguerite, but distanced herself from what she took to be the excessively academic and inappropriately unfeminine aspirations of women like Gournay (Green 2013). Scudéry's long novels and discussions of sensibility, in the moral conversations between men and women that make up the bulk of her works, provided an influential model of civilized society, and she, like other highly educated women at the period, saw herself as belonging to the republic of letters (de Scudéry 1654–1661, 1680, 1684, 1686, 1688, 1692; Pal 2012). Extending her literary correspondence, she attempted to incorporate into her literary acquaintance Elizabeth of Bohemia's friend, Anna van Schurman (Barthélemy and Kerviler 1878).

Scudéry's advocacy of polite salon society was taken up by Anne de Lambert, one of whose guests was Anne Dacier, an ikon of women's intellectual achievement at the time. Mary Astell hoped that English women would begin to emulate these French sisters, Scudéry and Dacier (Astell 1701; Dacier 1714; Lambert 1977). Her young friend, Mary Wortley Montagu, was particularly clearly influenced by Scudéry. During the early years of the eighteenth century, a whole phalanx of women writers emerged, who used novels, translations, and more hybrid genres, to enter into contemporary political debates, discuss the nature of virtue, and criticize contemporary marriage.

While questions might be raised as to whether Mary Delarivière Manley, Eliza Haywood, Sarah Fielding, Jane Collier, or Sarah Scott count as philosophers, they certainly explored ethical and political themes in their works, and aspired to influence the values and political attitudes of their contemporaries (Manley 1709; Carnell 2008; Haywood 1725; King 2012; Fielding 1744; Collier 1754; Scott 1762, 1766, 1772). The works of others, such as Catharine Trotter Cockburn clearly belong to standard philosophical genres, and her philosophical attitudes belong to the same Christian *eudaemonist* tradition accepted by these other women, a tradition which was developed in more radical directions by Catharine Macaulay, who, in turn, was a significant influence on Mary Wollstonecraft (Cockburn 1751; Green 2015; Hill 1995). This history of French and English women's ideas could be enriched with the addition of pathways of influence linking women from Germany and Russia, but for present purposes it is enough to recognize that we are in fact the beneficiaries of a rich legacy of women writers and thinkers, who have created values, transformed manners, and had an influence on religion. It is at least partly a result of this female intellectual legacy that we now inhabit the cultural milieu that we do.

When we begin to look at this alternative tradition, some names stand out as particularly influential, which are rather neglected if we only consider the women who dangle from the most prominent males. Christine de Pizan, Madeleine de Scudéry, and Catharine Macaulay play an exemplary role in the history of women's ideas, as original thinkers, who were not acolytes of illustrious men, but who were famous in their own right, and who have each left behind a large, rich corpus of works, that have been rather little studied as philosophy. A long time ago, Michèle le Doeuff coined the term 'Eloise complex' to describe the way in which women enter into philosophy as the lovers and acolytes of philosophical men. For these women 'their relationship to philosophy passes through their love for a man' (Le Dœuff 1977: p. 3). It now seems, that what appears to be, from Le Doeuff's point of view, an acceptance of philosophical subordination and failure of nerve on the part of women, is more properly seen as an artefact of the Hegelian history of ideas, which only admits women as danglers off the links in the philosophical chain of ideas, in virtue of their relationship to a male philosopher. The alternative genealogy has been suppressed because to see it one has to overcome the stifling legacy of Hegel and other male historians of the philosophical tradition.

So far, in this paper, I have been critical of the Hegelianism that Beauvoir's account embodies, but I want to end on something of a conciliatory note. Hegel's account of the dialectic of consciousness operates at two levels, the individual, and the historical. At the individual level, some descendants of his views have been very fruitful as the basis for accounts of the development of self-consciousness. That the first elements of self-consciousness involve recognition of the reflection of the self in the mirror, as not just as another phenomenal object, but as an image of the self, is now widely accepted in ethology and cognitive science. At least the germ of this idea is found in Hegel. Moreover, the social aspects, which flow from seeing oneself as an object for others, meditated on, for instance, in Beauvoir's *She Came to Stay* and in Sartre's chapters on concrete relations with others, also speak to our individual experience (Beauvoir 1995; Sartre 1943, 1983). But even at the individual level, the Hegelian conflict between consciousnesses, which emerges from the fact that we partly experience ourselves as objects defined by others, is arguably overdrawn. Very early on in human cognitive development, normal children develop what is called, 'a theory of mind' (Gopnik and Meltzoff 1997: pp. 143–60). Building on an innate capacity to imitate the actions of other humans, they begin to be able to recognize that the world, as experienced by another, is not exactly the same as the world as they are experiencing it. Early on we begin to recognize the other's subjectivity. We are able to put ourselves in another's shoes, and this does not require the unfolding of a historical dialectic. It is closely related to what is called 'sympathy' and is, in fact, a cognitive capacity that is necessary for the comprehension of utterances and the interpretation of actions and beliefs.

At the historical level, Hegel's teleological unfolding of spirit has, at first glance, even less going for it, than does his theory of the development of individual self-consciousness. Indeed, officially, even Sartre and Beauvoir reject it. From her first reading of Hegel she called his account of the absorption of individual consciousness into the Absolute, unfolding spirit, merely a 'consoling myth' (Green and Roffey 2010: p. 377). From the existentialist point of view, each individual consciousness is thrown into a particular historical moment and must deal with contingent being-in-itself, objectification by others, and congealed historical value, on their own. There is no guarantee of progress, but a repetition of conflict among embodied consciousnesses, who are ambiguously both object and subject. In bad faith, we refuse to recognize this fundamental ambiguity, and attempt to flee, either into pure transcendence and denial of facticity, or into passive objecthood and escape from responsibility.

Nevertheless, there are places where Beauvoir is tempted by the optimistic promise of a transposed dialectic. For Hegel, slave consciousness, although it is a 'dependent consciousness' is able to recognize the independent consciousness of the master, as the being that determines the reality of the slave's life. Through historical conflict and reversal, through revolution and the overthrow of the master, there can be progress to full self-consciousness, and mutual recognition, mediated by the state. If the relationship of man to woman, in fact, fits the relationship of the master to the slave, as Beauvoir claims, one would expect a parallel dialectic, a possibility of mutual recognition, as emphasized in Lundgren-Gothlin's interpretation of Beauvoir (Lundgren-Gothlin 1996: pp. 79–80, 248). And indeed, Beauvoir's account of pre-history suggests a dialectical unfolding of man's relationship with nature, that expresses itself in changes to his mythic characterization of woman. Near the end of *The Second Sex,* she also follows this logic, suggesting a future escape from bad faith and oppression, through the mutual recognition by the sexes of their ambiguous being as both flesh and spirit, nature and culture, object and subject, facticity and transcendence (Beauvoir 1949: p. 2:651).

Irigaray implicitly develops this element of Beauvoir, in her *Ethics of Sexual Difference* when she speaks of women's need to develop a female ethics with both a vertical mother–daughter dimension, and a horizontal sisterly dimension (Irigaray 1993: pp. 108–109). This she sees as necessary in order to overcome the suppression of all-female genealogy by the father–son relationship. For, women's Antigone-like burial has condemned her to providing the mute horizontal ground for the elevation of the male. Yet, without a history of women's actual philosophical contributions, Irigaray falls back on the material image of mucous, as representing something that has never existed and is already present, experienced in 'the prenatal and loving night known by both sexes' (Irigaray 1993, 109). Like Beauvoir, it seems that Irigaray is trapped within the Hegelian logic, within which women's world is biological and inarticulate. I would like to suggest, by contrast, that both reciprocity and the articulation of female ethics is impossible without a reconceptualization of the history of philosophy. Individual women are thrown into a moment of history and must deal with it each in her own way. In so far as the philosophy that they are taught continues to be the Hegelian history of men's ideas, the best they can do is react, by either absorbing the tradition and becoming acolytes and future danglers, or by repeating earlier women's rejection of men's ideas, or possibly by abandoning philosophy itself. Without philosophical mothers, we are orphans, who have no recognizable sisters. Those who remain within the cannon can dream of being the one, the exceptional female, who makes it as a philosopher and becomes that unique pearl in the male chain. But, if *we* don't recognize our foremothers, how can we expect the future to care about us? Only if there are, in the history of women's ideas, already female philosophers, whose ideas, values, and morals are worthy of study, can we expect that in the future our ideas will be remembered.

Yet, it has to be confessed that, from the point of view of the present, studying the history of women philosophers can be frustrating and might seem a lost cause, for, in the light of the continued dominance of a Hegelian understanding of the unfolding of spirit, it still offers two not very attractive options. One is to become an expert on one or more of the danglers, which requires immersion in the male stream from which they run as minor tributaries. The other is to immerse oneself in an alternative history of women's ideas, which, like women's studies in general, most men choose to ignore. What we need, as an alternative, is a cultural double helix, a sophisticated history in which we recognize both the evolution and development of men's ideas and the evolution and development of women's ideas, as well as the complex interaction between them. Already, through the study of our female cultural mitochondria, one begins to see the influence of the female line on the male. Critiques of monarchy inherit traits from the critique of marriage. Philosophy's turn towards sympathy and sentiment, during the eighteenth century, follows the rise of the novel of sensibility in the seventeenth. Rousseau as a child was immersed in the novels of Scudéry (Green 2009). When we, as a culture, recognize that the history of men's ideas has been intertwined with the history of the development of women's ideas, that cultural transmission develops along two intertwined axes, we will be beginning to overcome at the historical level the myth, according to which men are the only source of an active intellectual principle, the biological version of which has been completely discredited at the genetic level. No doubt this new self-conception of a sexed consciousness, responsible for the formation of a polity in which both sexes are equal citizens, will be mediated by a state in which there is parity. There is a paradox here. For the recognition that Beauvoir imagined might obtain between women and men, will only be possible through the transcendence of the Hegelian history of philosophy, which offered her the enormously productive proposition, that woman is the other of man, and has fostered a rich harvest of feminist philosophy. The Hegelian history of ideas, indeed, provided the ladder. It is time to kick it away.

References

Aristotle. (1941). *The basic works of Aristotle*. New York: Random House.

Astell, M. (1701). *A serious proposal to the ladies, for the advancement of their true and greatest interest. By a lover of her sex* (4th edn.). London: Printed by J. R for R. Wilkin.

Astell, M. (1706). *Reflections upon marriage* (3rd edn.). London: R. Wilkin.

Barthélemy, E., & Kerviler, R. (Eds.). (1878). *Un Tournoi de trois pucelles en l'honneur de Jeanne d'Arc. Lettres inédites de Conrart, de Mlle de Scudéry et de Mlle du Moulin*. Paris: Alphonse Picard.

de Beauvoir, S. (1949). *Le deuxième sexe*. Paris: Gallimard.

de Beauvoir, S. (1990). *Lettres à Sartre*. Paris: Gallimard.

de Beauvoir, S. (1995). *She came to stay* (trans: Moyse Y. & Senhouse, R.). London: Flamingo.

Bell, D. (1983). *Daughters of the Dreaming*. Melbourne and Sydney: McPhee Gribble/George Allen and Unwin.

Broad, J., & Green, K. (2009). *A history of women's political thought in Europe, 1400–1700*. Cambridge: University Press.

Butler, J. (1986). Sex and gender in Simone de Beauvoir's *The Second Sex*. *Yale French Studies* (72), 35–49. Simone de Beauvoir: Witness to a Century.

Butler, J. (1990). *Gender trouble: Feminism and the subversion of identity*. New York/London: Routledge.

Carnell, R. (2008). *A political biography of Delariviere Manley*. London: Pickering & Chatto.

Cereta, L. (1997). *Collected letters of a renaissance feminist* (trans: Robin, D.). Chicago and London: The University of Chicago Press.

Cockburn, C. T. (1751). In T. Birch (Ed.), *The works of Mrs. Catharine Cockburn, theological, moral, dramatical and poetical*. London: J. and P. Knapton.

Collier, J. (1754). *The cry: A new dramatic fable*. London: R. and J. Dodsley.

Dacier, A. (1714). *Des Causes de la corruption du goust*. Paris: Aux Dépens de Rigaud Directeur de l'Imprimerie Royale.

Deslauriers, M. (1998). Sex and essence in Aristotle's *Metaphysics* and biology. In C. A. Freeland (Ed.), *Feminist interpretations of Aristotle* (pp. 138–167). University Park, Pennsylvania: Pennsylvania University Press.

Fielding, S. (1744). *The adventures of David Simple: Containing an account of his travels through the cities of London and Westminster, in search of a real friend*. London: A. Millar.

Fonte, M. (1997). *The worth of women* (trans: Cox, V.). Chicago and London: The University of Chicago Press.

Gopnik, A., & Meltzoff, A. N. (1997). *Words, thoughts, and theories*. Boston, MA: MIT Press.

le Jars de Gournay, M. (2002a). *Apology for the woman writing and other works* (trans: Hillman, R. & Quesnel, C.). Chicago: University of Chicago Press.

le Jars de Gournay, M. (2002b) *Oeuvres complètes*. Paris: Champion.

Green, K. (1989). Femininity and transcendence. *Australian Feminist Studies, 4*(10), 85–96.

Green, K. (2009). Madeleine de Scudéry on love and the emergence of the 'private sphere'. *History of Political Thought, 30*(2), 272–285.

Green, K. (2014). *A history of women's political thought in Europe, 1700–1800*. Cambridge: Cambridge University Press.

Green, K. (2015). A moral philosophy of their own? The moral and political thought of eighteenth-century British women. *The Monist, 98*(1), 89–101.

Green, K. (1999a). The other as another other. *Hypatia, 17*(4), 1–14.

Green, K. (1999b). Sartre and de Beauvoir on freedom and oppression. In J. Murphy (Ed.), *Feminist interpretations of Jean-Paul Sartre* (pp. 175–199). University Park, Pennsylvania: Pennsylvania State University Press.
Green, K. (1995). *The woman of reason*. Cambridge: Polity.
Green, K. (2013). Women's writing and the early modern genre wars. *Hypatia, 28*(3), 499–515.
Green, K., & Roffey, N. (2010). Women, Hegel, and recognition in *The Second Sex*. *Hypatia, 25*(2), 376–393.
Haakonssen, K. (Ed.). (2006). *The Cambridge history of eighteenth-century philosophy*. Cambridge: University Press.
Haywood, E. (1725). *Memoirs of a certain Island adjacent to the Kingdom of Utopia*. London.
Hicks, E. (Ed.). (1977). *Le Débat sur le Roman de la Rose*. Paris: Honoré Champion.
Hill, B. (1995). The links between Mary Wollstonecraft and Catharine Macaulay: New evidence. *Women's History Review, 4*(2), 177–192.
Irigaray, L. (1985a). *Speculum of the other woman* (trans: Gill, G.). Ithaca, NY: Cornell University Press.
Irigaray, L. (1985b). *This sex which is not one* (trans: Porter, C.). Ithaca: Cornell.
Irigaray, L. (1991). Equal or different? In M. Whitford (Ed.), *The irigaray reader* (pp. 30–33). Oxford: Basil Blackwell.
Irigaray, L. (1993). *An ethics of sexual difference* (trans: Burke, C. & Gill, G. C.). Ithaca: Cornell University Press.
King, K. R. (2012). *A political biography of Eliza Haywood*. London: Pickering and Chatto.
de Lambert, M. (Anne Thérèse de Marguenat de Courcelles). (1977). *Madame de Lambert, oeuvres*. Paris: Champion.
Le Dœuff, M. (1977). Women and philosophy. *Radical Philosophy, 17*, 2–11.
Lloyd, G. (1984). *The man of reason: 'Male' and 'Female' in western philosophy*. London: Methuen.
Lundgren-Gothlin, E. (1996). *Sex and existence: Simone de Beauvoir's 'The Second Sex'* (trans: Schenk, L.). Hanover: University Press of New England.
Manley, M. D. (1709). *Secret memoirs and manners of several persons of quality, of both sexes. From the New Atalantis an Island in the Mediteranean*. London: John Morphew.
Marinella, L. (1999). *The nobility and excellence of women, and the defects and vices of men* (trans: Dunhill, A.). Chicago: Chicago University Press.
de Navarre, M. (1999). *Heptaméron*. Genève: Librarie Droz.
Nogarola, I. (2004). *Complete writings* (trans: King, M. L. & Robin, D.). Chicago and London: The University of Chicago Press.
Paakinen, I. (2016). *Gender and defense of women in Christine de Pizan's thought*. University of Helsinki.
Pal, C. (2012). *Republic of women. Rethinking the republic of letters in the seventeenth century*. Cambridge: Cambridge University Press.
de Pizan, C. (1997). *La Città della dame* (trans: Caraffi, P.). Milano & Trento: Luni Editrice.
de Pizan, C. (1959). *Le Livre de la mutacion de fortune*. Paris: Éditions A & J Picard.
de Pizan, C. (2001). *Le Livre de l'advision Cristine*. Études Christiniennes. Paris: Champion.
Sanday, P. (1981). *Female power and male dominance*. Cambridge: Cambridge University Press.
Sartre, J.-P. (1943). *L'être et le néant*. Paris: Gallimard.
Sartre, J.-P. (1954). *Réflexions sur la Question Juive*. Paris: Gallimard.
Sartre, J.-P. (1983). *Being and nothingness* (trans: Barnes, H.). London: Routledge.
Scott, S. (1762). *A Description of Millenium hall and the country adjacent: Together with the characters of the inhabitants, and such historical anecdotes and reflections, as may excite in the reader proper sentiments of humanity, and lead the mind to the love of virtue*. London: J. Newberry.
Scott, S. (1766). *The history of George Ellison*. London: A Millar.
Scott, S. (1772). *The life of Theodore Agrippa D'Aubigné*. London: Edward and Charles Dilly.
de Scudéry, M. (1654–1661). *Clélie, histoire romaine: dédiée à Mademoiselle de Longueville*. Paris: Chez Augustin Courbé.

de Scudéry, M. (1680). *Conversations sur divers sujets*. Paris: Claude Barbin.
de Scudéry, M. (1684). *Conversations nouvelles sur divers sujets*. Paris: Claude Barbin.
de Scudéry, M. (1686). *La Morale du monde, Conversations morales*. Paris: Sur le Quay des Augustins.
de Scudéry, M. (1688). *Nouvelles Conversations de morale dédiées au Roy*. Paris: Chez la Veuve de Sebastien Marbre-Cramoisy.
de Scudéry, M. (1692). *Entretiens de Morale dediez au Roy*. Paris: Chez Jean Anisson.
Simons, M. A. (1999). Richard Wright, Simone de Beauvoir, and *The Second Sex*. In *Beauvoir and The Second Sex* (pp. 167–184). Lanham: Rowman and Littlefield.

Chapter 9
Simone de Beauvoir and *the "Lunacy Known as 'Philosophical System'"*

Tove Pettersen

Abstract Tove Pettersen explores why Simone de Beauvoir was been neglected as a philosopher, and also why she denied calling herself a philosopher. One important reason is, Pettersen argues, the collision between her gendered experiences and the male-dominated philosophical tradition. This antagonism compelled her to transcend the discipline's boundaries and developed what later has been viewed as an original philosophy, but it also alienated her from the philosophical mainstream. It propelled some methodological obstacles that can shed light both on why she was excluded from the history of philosophy, and also why she denied being part of this tradition. Studying the characteristics of her philosophy may also reveal challenges women still face when doing philosophy, as well as the prejudices and methodological obstacles preventing them to gain their rightful place in the history of philosophy.

9.1 Philosophical and Methodological Connections Between *Diary* and Later Works

Although Simone de Beauvoir (1908–1986) was renowned as a writer in her lifetime, it was only after her death in 1986 and the posthumous publications of her letters and notebooks in the 1990s—thanks to the research of feminist philosophers and other scholars—that the philosophical novelty of her thinking was revealed. The reception of Beauvoir as a mere disciple of Jean-Paul Sartre has been widespread (Østerberg 2005: pp. 273–278; Lübcke 2003), even feminists have interpreted Beauvoir's philosophy as an adjunct to Sartre's. It has been claimed that "the framing metaphysics" used in *The Second Sex* (1949) is "laid out by Sartre" (Nye in Bauer

I would like to thank the Center for Gender Research at the University of Oslo where I had the privileged of being a guest researcher when I completed this article, and to the participants at the Researcher's seminar in particular for their very generous and constructive comments and suggestions. Many thanks also to Dr. Andrea Duranti for helpful comments on an earlier version of this paper, and professor Sigridur Thorgeirsdottir for very valuable suggestions.

T. Pettersen (✉)
Centre for Gender Research, University of Oslo, P.O. Box 1040, Blindern, 0315 Oslo, Norway

S. Thorgeirsdottir and R. E. Hagengruber (eds.), *Methodological Reflections on Women's Contribution and Influence in the History of Philosophy*, Women in the History of Philosophy and Sciences 3, https://doi.org/10.1007/978-3-030-44421-1_9

2001: p. 15), that this work can be understood as a "wedding gift [where] she brings a singular confirmation of the validity of Sartrism" (LeDoeuff in Bauer 2001: p. 16), an "application of Sartre's 'phenomenology of interpersonal relationships'" (Raymond in Simons 1999: p. 185), and that it's "central claim ... presupposes Sartre's argument that 'existence precedes essence'" (Kruks in Simons 1999: p. 185). The fact that Beauvoir herself claimed that "Sartre was a philosopher, and me, I am not" (Simons 1999: p. 9) has indeed added to such an understanding of her works.

In this article, I want to explore why Beauvoir was neglected as a philosopher for so long, and also why she denied calling herself a philosopher. Although there undoubtedly are several reasons working in conjunction, here I will concentrate on one in particular. As I see it, it was the collision between her gendered experiences and the male-dominated philosophical tradition that compelled her to transcend the boundaries of the discipline and carry out what later has been viewed as an original philosophy, that also alienated her from the philosophical mainstream. The antagonism between male philosophy and female experience propelled some methodological obstacles that might shed light both on why she was excluded from the history of philosophy, and also why she publicly refused to be part of this tradition.

In order to sustain this argument, I will consider Beauvoir's reflections on herself as well as her thoughts about philosophy as they appear in her diary for the years 1926–27, when she was an 18-year-old philosophy student. Not only does *Diary of a Philosophy Student* (de Beauvoir 2006) reveal Beauvoir's strong commitment to philosophy. As her relationship with Sartre started in 1929, *Diary* also proves Beauvoir's ownership of the most central ideas in her philosophy. I will discuss how many of these early ideas matured and later came to serve as part of her overall philosophy, and also how they inform her view on traditional philosophy. I shall then draw attention to her comprehension of philosophy as well as her novel philosophical method and ponder how they are possible explanations for her own—and others—claim of not being a philosopher.

As a woman, Beauvoir was confined by the dominant philosophical tradition based on male experience. However, as she in her works was in no mind to forfeit philosophy nor her gendered subjectivity and experience, her only option was to transcend several traditional philosophical concepts, definitions, and methods which together made it difficult to philosophize from the perspective of her own (female) condition. We find in *Diary* the seeds of Beauvoir's determination to analyze lived experience and to make philosophy capable of doing so. It is this view that both challenged traditional philosophy and opened up an entirely new field; feminist philosophy.

9.2 Women—Philosophy's Second Sex

Beauvoir was attracted to intellectual work early in her life (de Beauvoir 1963: pp. 21–22). By 1925, at the age of 17, she had decided she wanted to study philosophy at the Sorbonne. Her ambition not only defied some of the most entrenched notions of womanhood on which she had been brought up, but it was also at odds with her family's expectations. As far as her father was concerned, she was "denying her sex" in pursuing a career. "Philosophy", according to her mother, "mortally corroded the soul" (Simons 2006a: pp. 32, 39). In *Diary*, Beauvoir describes the conflict she also experienced between her passion for Jacques Champigneulle—her cousin whom she believed she would marry—and her desire to study. The self-sacrifice she thought would be required when married was impossible to reconcile with studying and living an autonomous life: "[H]e will simply introduce me to his life, but nothing will be changed. As for me, I am gambling all of myself!" (de Beauvoir 2006: p. 135). As to philosophy, Jacques "dismisses [it] with a smile", he sees it as a vain search (Ibid. 247).

Regarding women's capacity for thinking independently, she found few models or support in contemporary films, books, and art that interested her. Women were mostly depicted as defective men with an inferior capacity for rationality, morality, abstraction, and self-control, or as goddess-like, self-sacrificing creatures (Klaw 2006: pp. 13–14).[1] In philosophy, this view of women was particularly prominent. An outstanding feature of how the concept of reason is constructed in Western philosophy is its intention to transcend whatever is considered "feminine" (Lloyd 1984; Held 1993).[2] As a woman, Beauvoir belongs to the category of humans described and associated in the history of philosophy with the opposite of what it takes to be a philosopher: less rational, more emotional, less objective, more subjective, etc. By 1926, she already knows that pursuing her intellectual aspirations will not be without cost. "To feel eighteen years old, to know where happiness lies, and to tell yourself that you will probably never attain it; how colorless everything appeared" (de Beauvoir 2006: p. 71). She envisages a future of hard work and solitude. Despite internalizing some of the contemporary religious and cultural images of women,[3] with little support from her family, and despite the philosophical tradition that in sum doubted women's ability to do philosophy, in 1927 Beauvoir passed her exams in Greek, general philosophy and logic. In 1929 she obtained the prestigious *agrégation* in philosophy "and thus became the youngest philosophy teacher in France" (Musset 2010).[4]

---

[1]For references to such works, see Barbara Klaw, "The Literary and Historical Context of Beauvoir's Early Writings: 1926–27", in de Beauvoir (2006: pp. 7–28). As to the philosopher who possibly influenced her early philosophy, see Simons (1999: pp. 190–209).

[2]Both Lloyd (1984) and Held (1993) trace and analyze how, in the history of philosophy, the feminine has been excluded from influential philosopher's understanding of reason. Lloyd (1984: p. 2) puts it this way: "From the beginnings of philosophical thought, femaleness was symbolically associated with what Reason supposedly left behind–the dark powers of the earth goddesses, immersion in unknown forces associated with mysterious female power".

[3]For instance, Beauvoir reveals in *Diary* her internal struggle to gain control over herself and her life and suggesting that her sometimes failing willpower and nervousness might be rooted in the fact that she is woman (de Beauvoir 2006: p. 72).

[4]In France, university education was not equally open to women and men before 1924. Beauvoir took the *baccalauréat* exams in philosophy and mathematics in 1926, obtained in 1927 the Certificates in History of Philosophy, General Philosophy, Greek, and Logic, and in Ethics, Sociology, and Psychology in 1928. She wrote her graduate *diplôme* on Leibniz, and passed the philosophy agrégation in 1929 (Musset 2010).

In addition to revealing the conflict between Beauvoir's ambitions and her surroundings, *Diary* also shows great confidence in her own reflective and intellectual capacity. She believes in her ability to consider and analyze everything exhaustively. Indeed, this capacity she believes to be an important and dominating feature of herself, depriving her of spontaneity:

> I believe that this is a tendency that brings me a lot of pain and fatigue, but it is also my greatest claim to honor. I do not consider it to be at all indispensable to every being; I imagine exquisite natures that would not possess it, but me without it, I would be nothing. (de Beauvoir 2006: p. 124)

She actually instructs herself "to have faith in these truths that reflection has revealed to me and to have the morale [*courage*] to act on consequence" (Ibid. 55). When she comes to write *The Second Sex*, Beauvoir transforms her faith in her own point of view into advice for women who want to engage with literature, art, and philosophy: have faith in your own voice. Unfortunately, women often don't:

> [W]omen are still astonished and flattered to be accepted into the world of thinking and art, a masculine world; the woman watches her manners; she does not dare to irritate, explore, explode; she thinks she has to excuse her literary pretensions by her modesty and good taste; she relies on the proven values of conformism [...]. (de Beauvoir 2011: p. 745)

The consequences of such diffidence can be disastrous. "Crushed by respect for those in authority and the weight of erudition, her vision blocked by blinkers, the overly conscientious female student kills her critical sense and even her intelligence" (Ibid. 738). The point is that women do not by nature lack originality, but by nurture lack the bravery to use it. To achieve great things as an artist, Beauvoir writes in *Women and Creativity* (1966), one must challenge and oppose the world. Few women have the courage to do so—and those that do pay a high price (de Beauvoir 1987: p. 29). For Beauvoir, originality inheres in the individual's performance of their experiences and articulation of their subjectivity—either within the discipline if possible, or when used to press the discipline's boundaries. In order to be original, one must have something unique to contribute. Uniqueness is linked to one's subjectivity. Hence, by denying one's subjectivity and experience, one also denies the uniqueness necessary to be original. The only option left is to be a follower. One can have an "honorable [career] with such methods, but will not accomplish great things" (de Beauvoir 2011: p. 740). In *The Second Sex*, Beauvoir also advises the female philosophy student to resist this "nice girl" norm:

> The woman who chooses to reason, to express herself using masculine techniques, will do her best to stifle an originality she distrusts; like a female student, she will be assiduous and pedantic; she will imitate rigor and vigour. She may become an excellent theoretician and a reliable scholar; but she will make herself repudiate everything in her that is "different". (Ibid. 745)

Some of Beauvoir's originality can be explained by the fact that she takes her own experience of otherness seriously enough to subject it to philosophical analysis.

Instead of denying or ignoring that women are perceived as the Other, Beauvoir is reflecting over this fact when she writes her diary. In *The Second Sex* she subjects these reflections to a fully-fledged philosophical analysis, asking not only why, but also what it means for women to be the Other.[5]

9.3 The Philosophical Questions

In *Diary,* Beauvoir is determined to describe and analyze her moods and emotions as well as her own experiences (de Beauvoir 2006: p. 57). Personal and concrete experience must be taken as seriously as any abstract theory. Every phenomenon, she says, must be described from several different angles; how it was experienced, what caused this particular emotional response, how it might be understood and interpreted by others. As an 18-year-old philosophy student, she is already committed to this hallmark feature of existential philosophy, a feature which later becomes a prominent characteristic of her philosophy. In *Memoirs* she comments on the pleasure she took from studying Kant's *Critique,* but then she immediately adds: "Yet if it failed to explain the mystery of the universe and of my own existence, I really didn't know what could be the point of philosophy" (de Beauvoir 1963: p. 223).

Beauvoir applies her ability to scrutinize and analyze not only to theories, literature, and art, but also to her own situation, emotions, observations, and reactions. Writing a diary is her adolescent way of self-reflection, while the use of her own experience later becomes a basis for philosophical reflection, for instance in *She Came to Stay* (1943), *The Second Sex* and *On Aging* (1970). It has since become a characteristic feature of much contemporary feminist philosophy as well; women's experiences are subjected to a philosophical inquiry.

Beauvoir's *Diary* also contains her reflections on the conflict between self and others, and between autonomy and having close relationships (particularly with Jacques). This can be read as anticipating one of the main issues in her philosophy, the relationship between free and autonomous persons. The *Diary* also touches on her thoughts about love and commitment. As already mentioned, the adolescent Beauvoir struggles to balance the conventional image of women's love as self-sacrificing with her strong desire to study and determine her own life's path: "I would willingly consent to all sacrifice for a being I loved, but I would not want to exist only through him […]" (de Beauvoir 2006: p. 77). In addition to the conflict between interdependence and independence, *Diary* also seems to anticipate her later thoughts on how this conflict can be solved through an authentic, as opposed to inauthentic, love

[5]Beauvoir is not the only philosopher to make use of a concept of the Other. Hegel, Sartre and Lévinas do, too. But unlike most male philosophers, Beauvoir has first-hand experience of what it is to be the Other. At the same time, she has the capacity, education and possibility to philosophically analyze the experience. Beauvoir is thus situated both inside and outside this category. This is one reason why she can later give us a unique analysis of the lived experience of the Other.

(Petterson 2017). She wants a love, she writes in *Diary*, which "… is not a subordination, and [which] leaves the one who loves the care of seeking his own directions, of leading an independent, individual life" (Ibid. 76). She has, she explains, a powerful urge to "be herself", not what others want her to be, but she is also desperately longing for a kindred spirit (Ibid. 92). What she wishes for is a relationship that brings her love while allowing her to continue with philosophy. But living "through" another person would be impossible. "I have to have a goal to attain, a task to fulfill. I will never be able to be satisfied with what satisfies him or then out of complacence I will deceive myself, and preferring his esteem and affection to my own I will internally scorn myself—I will perhaps hold this against him" (Ibid. 134).

The pursuit of a freely chosen goal is a major topic of Beauvoir's philosophy, and in *The Second Sex,* she says women who forfeit their own goals for the sake of a husband are guilty of inauthentic love. Authentic love must be "founded on mutual recognition of two liberties" (de Beauvoir 2011: p. 520). In order to achieve such a relationship, it is mandatory for a woman to learn what Henrik Ibsen's Nora in the play *The Doll's House* (1879) finally learned, namely "that before being able to be a wife and a mother, she has to become a person", Beauvoir writes (Ibid.).[6]

It is tempting also to read many of Beauvoir's reflections in *Diary* on how important it is to assume responsibility for defining our own goals if our lives are to have any meaning as prefiguring central premises in the moral philosophy she published approximately 15 years later.[7] Her focus on having, or not having, a freely chosen aim in life also contains the seeds of her philosophy on transcendence and immanence. Furthermore, her fear of deceiving herself in *Diary* might pre-echo what she terms "mystification" in *The Ethics of Ambiguity* (1976: p. 98) and "bad faith" later on in *The Second Sex* (2011: p. 15). Women who believe it is sufficient to live through their husbands or children are suffering from "bad faith". It should be mentioned that with the publication of *The Ethics of Ambiguity*, Beauvoir is the only one of her contemporaries who published an existentialist ethics, in which she proves the critics wrong; existential ethics *is* possible (Pettersen 2015).

In *Diary*, she aspires to honesty and sincerity in her analyses and in her life. She is preoccupied with the split she experiences between her inner life and the external world. She has two selves, she points out, and there is a conflict between "a soul who would like *to be* and who must resign itself to appearing […]" (de Beauvoir 2006: p. 84). Beauvoir seems to be defining in her diary a posture she later terms "authenticity". She rejects the tendency to act and think in certain ways just because others do, i.e., what she later describes as living an inauthentic life (Ibid. 104).

> I must walk directly and energetically in my own direction. Live according to me, not according to others. I am *myself.* I am wrong, in determining my acts, not to act directly according to myself, to worry about the values established by others that are legitimate for them and not for me. (Ibid. 255)

---

[6]These are Ibsen's words: "HELMER: Before all else, you are a wife and a mother. NORA: I don't believe that any longer. I believe that before all else I am a reasonable human being, just as you are--or, at all events, that I must try and become one".

[7]For an elaboration on Beauvoir's concept of freedom and the necessity of creating meaning in our own lives, see Pettersen (2015: pp. 69–91).

9.4 Beauvoir's Philosophical Methods

In addition to the fact that many of the issues Beauvoir addresses in her diary are comparable to key philosophical ideas we find in her later works, she also seems to anticipate her preferred philosophical methods as a young philosophy student. Retrospectively, we can see that the approach she follows in *Diary* and develops more thoroughly later, basically involves first a description of the phenomenon as it appears or is experienced by her, followed by an analysis of the described phenomenon.[8] In her analysis, she uses both the theories of others and (in later years) her own analytical concepts and theories. Central is her phenomenological–existentialist framework with its emphasis on freedom, ambiguity, and authenticity. Finally, with her analysis as the foundation, she draws conclusions about human existence, about being in the world and about interaction (Holvek 1995: p. 73). The approach she calls for in *Diary* will loom large in later works: In *The Second Sex* and *The Coming of Age* (1970) she describes what it means to be "woman" and "older adult" from a variety of viewpoints; how it is experienced, how it can be explained from the perspective of biology, psychology, historic-materialism, cultures, etc. and finally how it can be evaluated: Women are the Other (de Beauvoir 2011), old people are "dehumanized" and treated "as outcast" (de Beauvoir 1972: pp. 8, 16). Her descriptions and approaches, we may add, are not only phenomenological, they are also *interdisciplinary*, and her conclusions are *normative* and founded in her existentialist ethics.

However, in *Diary* this approach is not portrayed as a deliberately chosen philosophical method, but rather a strong inclination to examine her experiences and observations in this way. Every phenomenon, she says, must be described from several different angles; how it was experienced, what caused a particular emotional response, how it might be understood and interpreted by others. She explains how "nothing ever sleeps in me", "everything remains present for me, on a sentimental order, as well as on the intellectual order" (de Beauvoir 2006: p. 124). She wants to "exhaust everything, to consider everything from all angles" (Ibid.). In *Diary* she instructs herself to keep a daily journal "with precise descriptions of what I have done, seen, thought"—while also spending time studying the work of others (Ibid. 275). Observing and analyzing emotions, moods, and behavior from a different perspective as they are displayed in different contexts is something she continues to do also in her later novels, including *She Came to Stay, The Second Sex, The Mandarins* (1954) and *A Very Easy Death* (1965).

When she meets conflicting viewpoints and claims, she finds reasonable points in most perspectives and arguments. But she refuses to privilege one perspective over another; she refuses to reconcile or reduce one to the other. Neither reality nor theory allows for simplistic solutions. In *Diary*, as in her mature work, she aspires to preserve the complexity and ambiguity she experiences in her own life. Acknowledging the ambiguity of human existence actually becomes an important ontological as well as

---

[8] Phenomenology, writes Vintges (1995: p. 47) restore the childlike approach to the world in adults by allowing for transcending the distinction between the subjectivity and the objectivity, for taking one's own presence in the world seriously. Beauvoir, it appears, never lost this approach.

methodological point of departure for her existentialist ethics. Beauvoir begins the *Ethics of Ambiguity* by depicting the apparently paradoxical ambivalence of human existence and its implications for philosophy. “As long as there have been men who live, they have all experienced this tragic ambiguity of their condition, but as long as there have been philosophers who think, most of them have tried to mask it” (de Beauvoir 2004: p. 290). For Beauvoir, existentialism is the only philosophical tradition that truly acknowledges the complexity of human existence (Ibid. 291), and she seems to have been in agreement with this existentialist premise by as early as 1926–27. Indeed, when existence is understood as ambiguous, it is difficult to see how a firm and pre-established principle could tell the whole truth about reality, let alone provide an all-inclusive guide for people’s actions.

Beauvoir’s *Diary* also demonstrates her nascent disbelief in universal principles, in a uniform method and an exhaustive truth. These insights would later become a cornerstone of her existentialist ethics. “No law can be imposed from the outside onto any of us. Those who depend on exterior principles are ‘barbarians’” Beauvoir writes in *Diary* (2006: p. 255), prefiguring her later portrayal of “the sub-man” in *The Ethics of Ambiguity* (1976: p. 42). In this book, she also explores the danger that follows devoting oneself to an external principle, a religion, or another human being, instead of taking responsibility for one’s own choices.

Beauvoir’s dismissal of universal principles does not only result from the fear that individuals will inflict harm and forfeit their own judgments and responsibilities in order to follow a principle. Additionally, it is linked to her epistemic position, namely that the production of knowledge is situated. Universal principles, with the intention of speaking on behalf of everyone, are deceptive. Or, as she writes in *Pyrrhus and Cineas* (1944), “One cannot have a point of view other than [one’s] own” (de Beauvoir 2004: p. 112). However, her rejection of universalism does not entail full-fledged relativism. Even if one only can speak on behalf of oneself, general aspects can be extracted from one’s own experiences. Let me enlarge:

For Beauvoir, individuals can both be uniquely and similarly situated. There are, for instance, no gender essences allowing one to universalize about all women (or all men). In addition, individual women (and men) live very different lives. However, having a female body and being situated in cultures where women are perceived as the Other, gives many women some shared experiences *qua* women. The experiences articulated, analyzed and evaluated in *The Second Sex*, which so many readers could relate to, are exactly of such a character. With this epistemological standpoint, Beauvoir challenges what traditionally is taken as universal without becoming a relativist. She develops a way *between* the claim that her theories are valid for everybody and valid for nobody but herself. Her analysis of her own lived experience might be valid for some; those similarly situated.[9] Consequently, Beauvoir’s philosophy challenges the traditional methodology of philosophy, theorizing from an alleged objective, contextless and universally valid perspective. But the apparently universal explanations

[9]Or, phrased differently, she indirectly argues that all universal claims are relative in the sense that they are relative to a time, place, culture, etc. However, this does not undermine the claims as such, it only confines the scope of their validity.

and analyzes of philosophy are incapable of capturing *her* life. The traditional voice of philosophy is the voice of men. This is indeed a paradox; the voice of philosophy, which aspires to talk and think on behalf of humankind, is inadequate when it comes to analyzing the female condition. Beauvoir's "One cannot have a point of view other than [one's] own" (de Beauvoir 2004: p. 112) disputes "the point of view from nowhere" and denies the representativeness of mainstream philosophy. By asking the critical question who speaks on behalf of whom, it is tempting to suggest that she anticipates not only feminist philosophy but also is open to philosophy of color, queer and postcolonial theory.

The voice of philosophy, aspiring to talk and think on behalf of humankind, falls short in the analysis of women's situation due to its male bias. For female philosophers, one way out is to ignore the existence of a female condition or claim that gendered reality can be transcended. Beauvoir's *Diary* reveals that neither are viable solutions in her opinion; she will neither consign her gendered experiences to oblivion nor judge them as insignificant to her philosophizing. "Clearly, no woman can claim without bad faith to be situated beyond her sex," she writes in *The Second Sex* (de Beauvoir 2011: p. 4), but the determination to find a way out of this predicament is revealed already in *Diary* where she is convinced that "pure analysis is foolishness" (de Beauvoir 2006: p. 67). But having experienced the "joyous astonishment in noting that … the abstract formulas of philosophy begin to live when they are clarified by quotes that resituate them in the current of individual consciousness" (Ibid. 66), Beauvoir is determined to find a voice of her own. As to the philosophical work involved in doing so, these are her words: "I feel capable of carrying it out successfully."

9.5 Philosophy, Literature, and Art

In *Diary* Beauvoir also portrays the tension she experienced between emotion and reason manifested as a dual attraction to philosophy and literature. This split is later explicitly elaborated on in her essay "Literature and metaphysics" (1946). In *Diary*, where she is determined to scrutinize both philosophical theories and the art of writing, she becomes aware of the distance separating her compelling emotions, and her cool, analytical intellect. She finds a void between the inexpressible and personal on the one hand, and the clear and precise yet impersonal on the other (de Beauvoir 2006: p. 59). She is stunned by the differences between what is intellectually known and what is felt (Ibid. 62). Beauvoir does not doubt her analytical capacity or ability to frame rational arguments. But neither does she distrust her experiences and emotions. And it is exactly analyzing and clarifying ordinary experiences and emotions that Beauvoir sees as one of the main purposes of philosophy: "[T]o intervene in a region of the self that likes to remain obscure and unconscious" (Ibid. 59–67; Holveck 1995: p. 71).

The question is how to articulate and analyze these complex lived experiences in the precise and clear language of philosophy. Since she found traditional philosophy notably lacking when it came to addressing her own real-life experiences, Beauvoir turned to literature and art. What fascinates her about art is almost the opposite of what attracts her to philosophy. While philosophy explains matters intellectually and ascribing to itself a universal value, art suggests through feelings of affinity, highlighting the individual. Art appeals by non-verbal means to subjective and individual emotions; It "awakens an echo in me," she writes (2006: p. 66).

It appears as if Beauvoir sees literature as a middle way between art and philosophy, emotion and reason, and between the inner, unarticulated echo of recognition and the universal, insensitive persuasion of philosophy. Literature can articulate the non-verbal, visualized expressions of the artists, and at the same time make "the abstract formulas of philosophy begin to live" (Ibid.). The strength of philosophy, as Beauvoir sees it, is that it possesses the tools to conceptualize individual experiences in an abstract and universally accessible way. It goes "straight to essentials" to the "general significance of things rather than their singularities" (de Beauvoir 1963: p. 158). Nevertheless, philosophical discussions "remain in [a] vacuum." Philosophy makes her feel like she is witnessing logical constructions, something she takes to be the opposite of being in touch with real life.

The point of expressing oneself, be it through art, literature or philosophy, is, she appears to be saying, to conquer solitude and loneliness, to show that "others have felt like you" (de Beauvoir 2006: p. 64). Within the bounds of traditional philosophy, she is unable to express her own experiences. Still, she needs the tools of philosophy to analyze and grasp the universal aspects of her lived experiences. But she also needs literature in order to articulate them fully.[10]

Already in her first publication, *She Came to Stay* Beauvoir intended to overcome inadequacies of orthodox philosophical analysis: *She Came to Stay* is a philosophical novel where Beauvoir uses both abstract philosophical theories and literary means. In this debut novel, Beauvoir draws on her own lived and felt experiences, and her analysis of these experiences is performed in what she in *Diary* described as "the clear and precise language of philosophy". Moreover, the philosophical questions she ponders are crucial to existential philosophy and ethics. These are questions about the possible relations between free consciousnesses, about whether or not there are limits to what one must accept from others in the name of freedom, and about self-deception and the distinction between subject and object. The tension between reason and emotion is indeed an important topic; what ought one to do when emotions run counter to rational and well-argued ideals? *She Came to Stay* also reveals the reciprocity Beauvoir believes exists between life and philosophy.

One could, of course, point out that Beauvoir was not unique among philosophers in addressing lived experience. Augustine, Descartes, and Kierkegaard are examples of philosophers whose personal experience informs their philosophy. Beauvoir nevertheless takes yet another step beyond mainstream philosophy. It is not the worries

---

[10]In *The Prime of Life*, written 34 years after *Diary*, she confirms this need for literature in order to be able to express what she wanted, which was "to communicate the element of originality in my own experience. In order to do this successfully I knew it was literature towards which I must orientate myself" (de Beauvoir 1962: p. 221).

of privileged and sovereign men she analyzes, but the lives of ordinary and embodied women. Analyzing new topics, such as housework, pregnancy, and sexuality in *The Second Sex*, not only brings new questions into the discipline, it challenges the very nature of philosophy because these topics are linked and associated with the supposed opposites of rationality, such as emotions, the body and private and personal attachments. Unlike Augustine, Descartes, and Kierkegaard who also include personal experiences, Beauvoir is confronted with an additional problem when addressing these issues. She wants to philosophize about questions that have no tradition, concepts or acceptance in traditional philosophy. For Beauvoir's purpose, there are no ready-made or adequate philosophical concepts or models for analyzing the experience of motherhood, housework, abortion or lesbianism. Ordinary language carries unwanted and gendered connotations: "Emotions", "the private", and "dependency" are at odds with the rational, independent and autonomous (male) agent as depicted in traditional philosophy. In other words, there is a huge conceptual gap between the topics she wants to address and traditional philosophy. This is one reason why she needs to lean towards literature.[11]

The reception of *She Came to Stay*—and other work—illustrates some of challenges faced by women philosophers: Beauvoir submitted the manuscript to Gallimard in 1941; most of it was written before Sartre began working on *Being and Nothingness*. Still, until the mid-'90s *She Came to Stay* continued to be read as Beauvoir's application of Sartre's philosophical ideas (Fullbrooks in Simons 2006b: pp. 42–43). As to her philosophical method in this and other works, it was in fact used against her; rather than being an authentic philosopher she was perceived as a writer in need of ideas which, one believed, Sartre generously provided her with. The paradox here is that by thoroughly engaging with the (male) history of philosophy—usually understood as a necessary but not sufficient token of being a philosopher—one risks accusations of non-originality, of siphoning one's ideas from others. But if one deviates too much from the tradition and is too original, one risks philosophical exclusion.[12] When Beauvoir was entering into dialogue with male philosophers it was understood as a sign of philosophical subordination, while the unique path she developed was regarded as not being of any philosophical interests. Quite frankly, as a philosopher, she was damned if she did, and damned if she didn't.

[11]Looking to another discipline in order to conceptualize—or re-conceptualize—is nevertheless not uncommon: Aristotle looked to biology, Plato and Descartes borrowed models and concepts from mathematics, Kierkegaard from theology, and Hume from what today is psychology. Which discipline is relevant depends on the topic. In analyzing lived experience Beauvoir is interested in literary methods, and in literary descriptions of lived experiences as a source for her phenomenological analysis.

[12]Obviously, Beauvoir relates to and engages with Sartre's philosophy—as she does with Bergson's, Descarte's, Husserl's, Hegel's, Kant's and Merleau-Ponty's philosophies.

9.6 The "Lunacy Known as 'Philosophical System'"

In *The Prime of Life*, written approximately 35 years after *Diary*, Beauvoir is still at pains to distance herself from traditional philosophy; she actually characterizes philosophy as the "lunacy known as 'philosophical system'". In the same passage, she also makes clear that she does *not* regard herself as a philosopher. As in *Diary*, she acknowledges her unique ability to "grasp philosophical doctrines" and "the ease which [she] penetrated to the heart of a text". In this, she is Sartre's superior, she says in *The Prime of Life*. But this ability is the result of her "lack of originality" she continues. She had no ideas of her own, she explains, and therefore she found it so easy to understand others'. Sartre, on the other hand, was so filled with his own original thoughts that "he found great difficulty in jettisoning his own viewpoint and unreservedly adopting anyone else's" (de Beauvoir 1962: p. 221). Beauvoir repeats this view several times, for instance in an interview with Margaret Simons and Jessica Benjamin in 1979:

> …Sartre was a philosopher, and me, I am not; and I never really wanted to be a philosopher. I like philosophy very much, but I have not constructed a philosophical work. I constructed a literary work. I was interested in novels, in memoirs, in essays such as *The Second Sex*. But this is not philosophy. (Simons 1999: p. 9)

Readers of *The Prime of Life* and of her interviews have frequently—and not surprisingly—noticed and discussed Beauvoir's disavowal of her own philosophical ambitions, talent, and achievements, placing herself beneath Sartre in importance (Simons 1990, 1999; Fullbrook and Fullbrook 1995; Bauer 2001; Daigle 2015). In particular, after the publication of *Diary*, where Beauvoir's philosophical aspirations and ideas are revealed, Beauvoir's denial of her philosophical achievements is baffling. However, there is much more at stake in these passages than Beauvoir putting herself in the shadow of Sartre. Nor are her statements merely a reflection of "the great difficulty women confront in accurately evaluating their own work, especially within an antifeminist context or discipline" (Simons 1990: p. 2). Read in light of her reflections on philosophy in *Diary*, it is possible to glean other aspects from these passages. Her tone of voice in *The Prime of Life* is, I suggest, not humble with regard to Sartre. Rather, it could be read as sarcastic and critical toward a traditional way of doing philosophy—an approach applied *also* by Sartre. What Beauvoir here distances herself from is an entire tradition within philosophy, namely the abstract system building—which Sartre according to her participates in, and also the narrow definition of philosophy that follows from it: "I am not a philosopher in the sense that I am a creator of a system […]" she claims (Simons 1999: p. 93). But not only does she distance herself from it. In these passages she also—once again—takes a stand *against* this type of philosophy as it is out of touch with lived experience. Instead of admitting its linkage with reality and subjectivity, these abstract systems are presented as being built on universal truths established by one's favorite philosophical ancestors—along with one's own brilliant corrections of their mistakes. Moreover, by terming it "lunacy" she ridicules it. Her depiction of its outcome is not exactly one of admiration either. She wants to know "*how* certain individuals are capable of getting results from that conscious venture into lunacy known as a 'philosophical system', from which they derive that obsessional attitude which endows their tentative patterns with universal insight and applicability" (de Beauvoir 1962: p. 221). In *The Prime of Life* and in the interviews, as in *Diary*, Beauvoir clearly demonstrates her disdain for philosophers who simply engage with the ideas of others. "Expounding other people's beliefs, developing, judging, collating, and criticizing them—no, I failed to see the attraction of this". She is "wondering how anyone could bear to be someone else's follower, or disciple" (Ibid.). Beauvoir did not want to be one of those philosophers who, as Holvek puts it, "spent too much time on nothing" (1995: p. 69).

Beauvoir's comments on philosophy in *The Prime of Life* and in the interviews from 1979 and 1985 are remarkably consistent with her earliest views on philosophy written in 1926–7. In *Diary*, *The Prime of Life* and the interviews she takes a well-defined position towards the traditional endeavor of establishing an exhaustive philosophical theory, capable of revealing the entire truth or being universally applicable.[13] She seems not to have wanted to do philosophy this way. She has no confidence in a philosophy that is isolated from real life; those who engage in this activity become "obsessed", and they mistake their own provisional patterns as universal.[14]

Philosophy as "system building" does not stay in contact with concrete, individual experience—exactly what Beauvoir wants to philosophize over. This is obviously one reason why she never was attracted to the systematic and "pure" philosophy. Another reason might be that since philosophical system building mostly consists of confronting and puzzling together abstract ideas, and stripping phenomena of their sensuous and ambiguous dimensions, the type of originality Beauvoir encourages in *The Second Sex* and *Women and Creativity* is not required. Beauvoir repeatedly claims that original contributions spring from one's unique experience. When philosophers engage with ideas that are completely out of touch with their own life, few of them are anything but disciples. Such philosophers are most likely doing what she warns her students against—killing their critical sense and even their intelligence (de Beauvoir 2011: p. 738). In *The Prime of Life*, she implies that this is the kind of philosopher she never was—precisely what she seemed to earlier promise herself in *Diary*. She does not say it is impossible to be an original philosophical system builder, but "in this

[13]One might object that this type of philosophy was pretty much over after Hegel's last attempt at the creation of an encompassing philosophical system, i.e. by 1830. However, this tendency can still be found in moral philosophy (i.e. deontology and utilitarianism as universal systems of right and wrong), and political philosophy (i.e. Rawls' *Theory of Justice* and Nussbaum's list of capabilities; both intended to be universal).

[14]This is an idea conveyed in *The Ethics of Ambiguity* in her description of the "serious man", a potentially fanatic person who "mask[s] the movement by which he gives [values] to himself, like the mythomaniac who while reading a love-letter pretends to forget that she has sent it to herself" (de Beauvoir 1976: p. 47).

field a genuinely creative talent is so rare" (1962: p. 211).[15] Her own creative talent, in her view, lies in literature because it gives her space to voice lived experience.

When Beauvoir is evaluating her own work on the basis of the discipline's norms, they diverge sharply from the established standards. For Beauvoir, who has criticized these norms during her entire life, this is a *good* thing. Being out of sync with the restrictive tenets of traditional philosophy shows she has been true to the commitment she made and the critique she fashioned as a young woman in her journal. She was *never* a follower; she did *not* stifle her own creativity, subjectivity or critical thinking. By saying that she is not a philosopher, she positions herself where she wants to be: outside the philosophical establishment.[16]

When Beauvoir retrospectively depicts herself primarily as a writer and Sartre a philosopher (Simons 1999: p. 11), it is not, I suggest, because she *values* his works more than her own, nor because she wants him to be the first and subject herself to a position as the Other in their relationship. Nor is it because she holds traditional philosophy in higher regard than literature, or because she "disqualifies herself from the ranks of philosophers" (Daigle 2015: p. 18). Such explanations, focusing on Beauvoir's lack of philosophical self-esteem or her interpersonal relationships are not only countered by the reading of her *Diary*. Additionally, they also contribute to reducing enduring obstacles faced by many female philosophers—past and present—to an individual and isolated problem. Consequently, such explanations fail to account for *systematic* mechanisms excluding many women from the discipline of philosophy. In retrospective, it should be evident that the main reason Beauvoir was neglected as a philosopher for so long and denied calling herself a philosopher, was not simply a "personal problem". It was a structural problem.

Beauvoir did not want to be termed a philosopher based on how philosophy was measured by the traditional standards of philosophy as an institution. However, not wanting to be a philosophical system builder does not entail that she does not want to be a philosopher at all, or must present only incoherent and fragmented thoughts. Nor is she alone in having difficulties with the mold of traditional philosophy. Several influential twentieth century philosophers, such as Heidegger, Nietzsche, Wittgenstein, and Arendt kept a distance to philosophy and challenged the discipline. In hindsight, this contributed also to their originality. However, none of these "outsiders" were critical of philosophy because of its exclusion of women and the feminine. It is

[15]Sartre is one of the few to possess originality in this field, she claims. Beauvoir is here referring to Sartre's *Being and Nothingness*. However, scholars have demonstrated that it is very likely that Sartre based this work on Beauvoir's *She Came to Stay*, and some claim that it is the original source of Sartre's philosophical system (Fullbrook and Fullbrook 1995).

[16]One might object that even if what has been said so far might hold true for *The Second Sex* and *She Came to Stay*, it is not correct with regard to *Pyrrhus and Cineas* and *The Ethics of Ambiguity*. These texts are more traditional in style and form than *The Second Sex* and *She Came to Stay*. But these works too challenge traditional moral philosophy in profound ways, though it is impossible to discuss this further for reasons of space.

from the perspective of her contemporary *patriarchal* philosophical discourse that Beauvoir's work fell outside the boundaries.[17]

Perhaps it is only in retrospect, with feminist philosophy established, that we can fully grasp the novelty of Beauvoir's philosophy as well as understand the reception and fierce resistance against her, and many other female philosophers' work. Feminist philosophers have exposed the male bias of traditional philosophy with regard to topics and methods. They have de-masked the gender bias inherent in the philosophical concepts and categories, in ontology, epistemology and the history of philosophy and revealed how this has contributed to the exclusion of women (Lloyd 1984; Held 1993; O'Neill 1998). New methodological models and conceptual tools have been developed by feminist philosophers in order to grasp the mechanisms at play with regard to women in philosophy. For example, the reception of Beauvoir's work can now be understood as a case of what Miranda Fricker terms "testimonial injustice", i.e., where someone is discredited as a "knower" based on prejudices directed towards someone's social identity, such as being a woman (Fricker 2007: p. 54). Moreover, the lacuna in traditional philosophy's epistemic recourses, rendering women's lived experience incomprehensible for traditional philosophical analysis, might be understood as an example of "hermeneutical injustice". This epistemic defect aspired Beauvoir towards transgressing traditional philosophy. Both phenomena, i.e. testimonial and hermeneutic injustice, cause ethical as well as epistemic harm. In a context where rationality is strongly linked to being considered a human, one is degraded and humiliated *qua* human being by such testimonial discredit (Ibid. 44). The entire epistemic system is damaged, as the disregard and exclusion of women's experiences obstructs the circulation of critical ideas and limits the freedom of expression.

Karen Warren compares the traditional philosophical canon with a house; the foundation and framework are given. Some variations with regard to the interior are possible, but the house itself cannot be given another shape, form or location. Therefore, new questions, topics, and methods presented by feminist philosophers are hard to integrate. Often they challenge the construction of the house and are therefore left "outside" (Warren 2009: pp. 1–26). Beauvoir's philosophy, especially *The Second Sex,* poses a definite challenge to the foundation, framework, and location of the traditional "house of philosophy". This may be one important reason why she has been neglected so long, and why it were feminist philosophers who "(re)discovered" her work as philosophy.

Feminist philosophy gives us the tools to explain why Beauvoir created an innovative philosophy, and also why her work was not considered philosophy—by herself or others. It has enabled us to reinterpret the works of Beauvoir—and other female philosophers of the past. One can now re-categorize female thinkers and their works,

---

[17] It might be argued that no matter how critical Heidegger, Nietzsche, Wittgenstein and Arendt were towards traditional philosophy, none of them explicitly contributed to challenge the patriarchal aspects of philosophy. This holds true even if feminist philosophers later have found their works valuable exactly in this respect.

and reveal the structures that expelled them from the discipline (Gardner 2004).[18] Whether or not Beauvoir today would consider herself a feminist philosopher, we can only speculate.

9.7 Philosophy Out of Bounds

What are—in retrospect—the characteristics of Beauvoir's form of philosophy? To sum up, philosophy, according to Beauvoir, is not limited to constructing an exhaustive abstract system out of touch with real life, based on interpretations, critiques, and confrontations with the ideas of one's philosophical ancestors. It is not primarily to compose theories in competition with other theories concerned with accessing or determining the "truth." Transforming life's puzzles into coherent systems is not impossible, but Beauvoir reminds us that the resulting structures are inauthentic nonetheless. Lived experience is fractured and ambiguous; life is *not* as unequivocal, coherent and consistent as traditional philosophers often present it. Philosophers *create* the structures, coherence, connections, similarities, and differences, features manifesting the particular theorist's view on the matter—not the universal truth.

For Beauvoir philosophy is first and foremost to scrutinize selected and ambiguous fragments of life, i.e., to intellectually grasp, articulate and critically analyze lived experience. In order to do this, she confronts philosophical ideas with reality and experience rather than other ideas. Such philosophical analysis may "conquer solitude and loneliness", to show that "others have felt like you", as well as de-masking dominant narratives upholding the *status-quo*. This, in turn, may lead to action and changes. By revealing the dominant myths about women, men, love, old age and so on, Beauvoir demonstrates what the job of the philosopher is; "to re-describe how things are in a way that competes with the *status-quo* story and leave us craving social justice" (Bauer 2015: p. 16). Beauvoir's resistance to the dominant "grand narratives" is one hallmark of her philosophy and can be read as anticipating not only feminist philosophy but also post-structuralists' and postmodernists' criticism of traditional philosophy.

Beauvoir's commitment to philosophize on lived experience, and not to reduce or mask the complexity and ambiguity of the human condition, also caused her to cross some established boundaries. She famously declared that there is no strict border between life and philosophy; every human act is a philosophical choice (de Beauvoir 2004: pp. 217–18). She also saw the strict distinction between literature and philosophy as essentially false and used literary forms and techniques as part of her philosophical method. In her moral philosophy, she challenged the division between

[18] A parallel can be drawn here: some words, categories and distinctions have to be "invented" to be able to classify phenomena. Before feminists conceptualized "sexual harassment", "double day", "*ad feminam*", the terms were not there to label certain phenomena. But once conceptualized, many women felt their experience could be articulated, and could see that others shared their experiences. That was when one could start to theorize about them.

ethics and morals, and ethics and politics (Pettersen 2007, 2014). Transcending established demarcations within and between disciplines is another feature of Beauvoir's philosophy.

When traditional philosophy proved inadequate to Beauvoir's purpose, she developed her own approach, characterized by combining philosophical inquiry with the ability to bring the abstract to life by using literature as well as women's experiences. Using the experiences of ordinary women as examples and counter-examples capable of demonstrating points of general philosophical interests, and not merely being anecdotes, deeply challenges a tradition where only men have been recognized as valid representations of what is human. Beauvoir's philosophy is furthermore marked by the combination of what appear to be opposites: the abstract and the concrete, emotion, and reason, the universal and the particular, the objective and subjective. By transcending traditional binary logic and including ambiguity, she is anticipating contemporary feminist philosophy's attempt to develop anti-dichotomous concepts and models (Pettersen 2008, 2010). Most importantly, Beauvoir lifts women's experiences out of the private domain, acknowledging them and making them intellectually accessible, capable of demolishing conventional ideas and philosophical models that for millennia contributed to the deprivation of women's freedom. Far from rejecting philosophy, she transformed the discipline from within and contributed substantially to its development. Studying the characteristics of her philosophy also reveals the challenges women face when doing philosophy, as well as the prejudices and methodological obstacles preventing them to gain their rightful place in the history of philosophy.

References

Bauer, N. (2001). *Simone de Beauvoir, philosophy, and feminism*. New York: Columbia, University Press.
Bauer, N. (2015). *How to do things with pornography*. Cambridge, Massachusetts, London Harvard University Press.
de Beauvoir, S. (1949). *Le deuxième sexe*. Paris: Gallimard.
de Beauvoir, S. (1962). *The prime of life*. New York: Penguin books.
de Beauvoir, S. (1963). *Memoirs of a dutiful daughter*. Middlesex: Penguin books.
de Beauvoir, S. (1972). *The coming of age*. New York: Warner Paperback Library.
de Beauvoir, S. (1976). *The ethics of ambiguity*. New York: Citadel Press.
de Beauvoir, S. (1987). Women and creativity. In T. Moi (Ed.), *French feminist thought* (pp. 17–32). Oxford: Basil Blackwell.
de Beauvoir, S. (2004). *Simone de Beauvoir. Philosophical writings*. In M. A. Simons, M. B. Timmermann, & M. B. Mader (Eds.). Urbana and Chicago: University of Illinois Press.
de Beauvoir, S. (2006). *Diary of a philosophy student. Volume 1, 1926–27*. In B. Klaw, S. Le Bon de Beauvoir, M. A. Simons, M. Timmermann (pp. 29–50). Urbana and Chicago: University of Illinois Press.
de Beauvoir, S. (2011). *The second sex*. New York: Random House.
Daigle, C. (2015). Making the humanities meaningful. Beauvoir and the appeal. In T. Pettersen & A. Bjørsnøs (Eds.), *Simone de Beauvoir—A humanist thinker* (pp. 17–28). Leiden: Brill/Rodopi.
Fricker, M. (2007). *Epistemic injustice: Power and the ethics of knowing*. Oxford University Press.

Fullbrook, S, & Fullbrook, E. (1995). Sartre's secret key. In *Feminist interpretations of Simone de Beauvoir* (pp. 97–111). Pennsylvania: The Pennsylvania State University Press.
Gardner, C. V. (2004). *Women philosophers. Genre and boundaries of philosophy*. Boulder, Colorado: Westview Press.
Held, V. (1993). *Feminist morality: Transforming culture, society, and politics*. Chicago: University of Chicago Press.
Holveck, E. (1995). Can a woman be a philosopher? Reflections of a Beauvoirian housemaid. In M. A. Simons (Ed.), *Feminist interpretations of Simone de Beauvoir* (pp. 67–78). Pennsylvania: The Pennsylvania State University Press.
Ibsen, H. (1879). *A Doll's House*. Retrieved November 1, 2015, from http://www.classicreader.com/book/2011/4/.
Klaw, B. (2006). The literary and historical context of beauvoir's early writings: 1926–27. In B. Klaw, S. Le Bon de Beauvoir, M. A. Simons, & M. Timmermann (Eds.), *Diary of a Philosophy Student. Volume 1, 1926–27* (pp. 7–28). Urbana and Chicago: University of Illinois Press.
Lloyd, G. (1984). *The man of reason. 'Male' and 'female' in western philosophy*. Minneapolis: University of Minnesota Press.
Lübcke, P. (ed.). (2003). Fransk filosofi: Engagement og struktur. København: Politikkens Forlag A/S.
Musset, S. (2010). Simone de Beauvoir (1908–1986). In *The Internet encyclopedia of philosophy*, ISSN 2161-0002. Retrieved October 1, 2015, from http://www.iep.utm.edu/beauvoir/.
O'Neill, E. (1998). Disappearing ink: Early modern women philosophers and their fate in history. In J. A. Kournay (Ed.), *Philosophy in a feminist voice: Critiques and reconstructions* (pp. 17–33). Princeton: Princeton University Press.
Østerberg, D. (2005). *Jean-Paul Sartre*. Oslo: Gyldendal Norsk Forlag AS.
Pettersen, T. (2007). Freedom and feminism in Simone de Beauvoir's philosophy. *Simone de Beauvoir studies volume 24, 2007–2008* (pp. 57–65). Menlo Park California: Simone de Beauvoir Society.
Pettersen, T. (2008). La joie existentielle et l'angoisse dans la philosophie morale de Simone de Beauvoir. In J. Kristeva (Ed.), *(Re)découvrir l'œuvre de Simone de Beauvoir Du Deuxième Sexe à La cérémonie des adieux* (pp. 212–225). Paris: Le Bord de L'eau.
Pettersen, T. (2010). Acting for others: Moral ontology in Beauvoir's pyrrhus and cineas. In *Simone de Beauvoir studies volume 26, 2009–2010* (pp. 18–27).
Pettersen, T. (2015). Existential humanism and moral freedom in Beauvoir's ethics. In T. Pettersen & A. Bjørsnøs (Eds.), *Simone de Beauvoir–A humanist thinker* (pp.69–91). Leiden: Brill and Rodopi.
Pettersen, T. (2017). Love—according to Simone de Beauvoir. In L. Hengehold & N. Bauer (Eds.), *A Companion to Simone de Beauvoir* (pp. 160–173). Hoboken: Wiley Blackwell.
Simons, M. A. (1990). Sexism and philosophical canon: On reading Beauvoir's The second sex. *Journal of the History of Ideas*, *51*(3), 437–504.
Simons, M. A. (1999). Beauvoir interview (1979). In *Beauvoir and The second sex. Feminism, race, and the origins of existentialism*. Lanham: Rowman & Littlefield.
Simons, M. A. (2006a). Beauvoir's early philosophy: 1926–27. In B. Klaw, S. Le Bon de Beauvoir, M. A. Simons, & M. Timmermann (Eds.), *Diary of a philosophy student. Volume 1, 1926–27* (pp. 29–50). Urbana and Chicago: University of Illinois Press.
Simons, M. A. (Ed.). (2006b). *The philosophy of Simone de Beauvoir. Critical essays*. Blomington and Indianapolis: Indiana University Press.
Vintges, K. (1995). The second sex and philosophy. In M. A. Simons (Ed.), *Feminist interpretations of Simone de Beauvoir* (pp. 45–65). Pennsylvania: The Pennsylvania State University Press.
Warren, K. (Ed.). (2009). *An unconventional history of western philosophy: Conversations between men and women philosophers*. Lanham: Rowman & Littlefield.

Chapter 10
Arendt, Natality, and the Refugee Crisis

Robin May Schott

Abstract Arendt had little interest in the question of women in the philosophical canon. But her concept of natality and its links to reflections on difference, contestation, and multiplicity have struck a chord with many contemporary (and feminist) philosophers. Through her historical and theoretical examinations of the conditions for human life, political life, and ethical life—and for their transgressions—she has become a powerful intellectual figure. Her concept of natality, which focuses on the role of plurality, newness, and spontaneity for political life, offers distinctions and priorities that provide a necessary orientation for reflecting on the contemporary refugee crisis. In an Arendtian spirit, we must defend a conception of an open political sphere that is constantly renewed by newcomers, and that is not reducible to social or cultural identities, in order to protect refugees as obligated by both international conventions and moral relations.

10.1 Introduction

Hannah Arendt (1906–75), along with Simone de Beauvoir, is widely considered one of the most significant women thinkers in twentieth century Western philosophy. Indeed, in her life she experienced a long list of "firsts". She was the first woman to give the Christian Gauss seminars at Princeton University in 1953; the first woman to be appointed full professor at Princeton in 1959; one of the few female academicians to receive the Lessing Prize of the Free City of Hamburg in 1959, the Emerson–Thoreau Medal of the American Academy in 1969, and the Sigmund Freud Prize of the Deutsche Akademie für Sprache und Dichtung in 1967 (Dietz 1995, 18). And yet she threatened to refuse the Princeton appointment, because the university stressed the "first woman" aspect in their report to the *New York Times*. When interviewed about the Princeton appointment, Arendt responded, "I am not disturbed at all about being a woman professor, because I am quite used to being a woman" (Young-Bruehl

R. M. Schott (✉)
Danish Institute for International Studies, Østbanegade 117, 2100 Copenhagen, Denmark
e-mail: rms@diis.dk

S. Thorgeirsdottir and R. E. Hagengruber (eds.), *Methodological Reflections on Women's Contribution and Influence in the History of Philosophy*, Women in the History of Philosophy and Sciences 3, https://doi.org/10.1007/978-3-030-44421-1_10

1982, 272–3). She was keen to reject the role of "exception woman" just as she was highly critical of the position of "exception Jew"—the dangers of which she had learned from the complex history of the Jews in Germany.

Yet being a woman, like being a Jew, was a fact for Arendt—and to deny it would have been, as she noted, insane.[1] And indeed her biography reflects the facts of being a woman—starting with her struggles with her mother, Martha Arendt. Her husband Heinrich Blücher was bitter over her mother's lack of respect for her work, even though "she was the one who, more than any other timid blockhead in our circle, simply and thoughtlessly took you for a man" (Young-Bruehl 1982, 236–7). As a young student Hannah Arendt had a love affair with Martin Heidegger, a relation in which Heidegger thought of her for years as his Muse, but not his equal. And when Arendt sent him the German translation of *The Human Condition*, he met it with a burst of hostility. Arendt confided to Karl Jaspers in 1961, "I know that it is intolerable for him that my name appears in public, that I write books, etc. I have really fibbed to him about myself all the while, behaving as though none of this existed and as if I, so to speak, could not count to three, except when it came to giving an interpretation of his own things....But suddenly this fib became quite boring to me and I have paid for my change of mind with a knock on the nose" (Young-Bruehl 1982, 307). And her own advice to younger women was to urge women to independence; but her maxim was "*Vive la petite difference*" (Young-Bruehl 1982, 238)![2]

Arendt was highly critical of the women's movement, noting "the movement has never united to achieve concrete goals....The vain attempt to found a women's party shows the questionableness of the movement: it is the same questionableness as that of the youth movement, which is a movement only for youth as the women's movement is only for women. One is as abstract as the other" (Young-Bruehl 1982, 96). The past achievements of the suffrage movement had come about from political action, and insofar as the women's movement did not translate its ideology into concrete political goals, it remained abstract. Not only was Arendt critical of the lack of focus on the political arena in the women's movement, she also opposed any tendency to divorce women's issues from larger political concerns, just as she opposed divorcing Jewish issues from international and national political concerns. In this respect, Arendt's position followed that of Rosa Luxemburg, who argued that the oppression of women, like the oppression of Jews, would only come to an end with the advent of socialism. While Arendt did not follow Luxemburg in advocating socialism, she did believe that women's issues should be part of a larger political struggle, and that women should pursue concrete political goals such as legislation

[1]Her book on Rahel Varnhagen (Arendt 1957/1997) reflects her interest in exploring the questions of being a woman and a Jew in Germany.

[2]Bonnie Honig rejects the standard view that this phrase refers to sexual difference. She claims, "*La petite difference* is an intra-sex/gender difference" which characterized Rosa Luxemburg's difference from other women, "the refusal of membership, the choice of difference or distinction over a certain kind of equality (Honig 1995a, b, 151)."

for equal employment opportunities (Young-Bruehl 1982, 97).[3] However, after 1948 Arendt herself never took an active role in political life, and was skeptical about whether women should be political leaders (Young-Bruehl 1982, 238).

Just as she met the women's movement with skepticism, the women's movement has met Arendt with skepticism as well. The American feminist poet Adrienne Rich described *The Human Condition* as a "lofty and crippled book" that "embodies the tragedy of a female mind nourished on male ideology" (cited in Honig 1995a, b, 2). And the Scottish-Canadian feminist political theorist Mary O'Brien read Arendt as "a woman who accepts the normality and even the necessity of male supremacy" (cited in Dietz 1995, 24). Arendt's relationship with Heidegger has been an endless source of sexist comments, and has also contributed to the view that her philosophical method is heavily dependent on Heidegger. Indeed, Arendt is often chastised for the "missing body" in her work. Following the traces of ancient philosophy and the city-state, Arendt viewed the natural, biological body as an unquestioned condition of life, but decidedly not as a topic for philosophical life. And to the chagrin of many contemporary feminist philosophers, she treated the body as a genderless form of existence, leading Mary Dietz to remark, "[I]t is certainly curious that Arendt never makes this central feature of the human condition an integral part of her political analysis" (Dietz 1991, 240). In re-opening my old copy of *The Human Condition* I can also read the penciled handwriting in the margins from my days in graduate school. Where Arendt opens the discussion of labor as a biological process of the human body, I wrote "no feminist analysis". And Arendt's disinterest in the category of gender is accompanied by a disinterest in the concepts of maternity and femininity.

Mary Dietz distinguishes many of the feminist interpretations of Arendt as framing her as either a phallocentric or a gynocentric thinker. While Rich and O'Brien give voice to the phallocentric Arendt, other feminists approach her as a gynocentric thinker. The American feminist philosopher Nancy Hartsock, for example, wrote that "Arendt remains an interesting and important example whose work indicates some of the beneficial theoretical effects women's experience of both connection and individuation may have…." (cited in Dietz 1995, 27). Political theorist Jean Bethke Elshtain noted, Arendtian concepts such as natality "stir recognition of our own vulnerable beginnings and our necessary dependency on others, on (m)other" (ibid.). From both sides of the debate—the phallocentric Arendt or the gynocentric Arendt—the question of whether Arendt refuses to think as a woman, or thinks as a woman, has been decisive in feminist assessments.[4]

Subsequent to these initial debates in the 1970s and 1980s, feminists have moved beyond the phallocentric-gynocentric debates to bring Arendtian theory to bear on

---

[3]This view of women's issues aligns with her commitment to distinguish social questions from political questions, a view which she also maintained in her controversial 1959 essay on desegregation, "Reflections on Little Rock".

[4]Arendt thought there was nothing intrinsically masculine or feminine about the public realm; similarly, she would reject the view that thought places demands to think in a masculine or feminine way. In this respect, she would differ with Beauvoir's view that in common usage as well as in scientific thought, "man represents both the positive and the neutral…whereas woman represents only the negative, defined by limiting criteria, without reciprocity" (Beauvoir 1952/1974, xviii).

conceptual problems that are important for contemporary feminist theory (Dietz 1995, 33). Bonnie Honig focuses on Arendt's conception of political action as engendering identities. She reads Arendt's acting subject as a performative, self-creative, and transformative self, a self which is itself multiple and the site of struggle. Arendt's notion of agonism becomes in her view a resource to understand action as an event, a disruption of the ordinary course of things, and a site of resistance to the normalizing rules that govern behavior (Honig 1995a, b, 140–142). In her Arendtian inspired feminist theory, this notion of action and contestation becomes a resource for radicalizing Arendt's own theory and politicizing the public/private divide on which Arendt insisted.

Hence, feminist readings of Arendt follow patterns similar to feminist readings of male philosophers in the canon of Western philosophy. While the early interventions in re-interpreting the history of philosophy dismissed male philosophers as patriarchal (a judgement echoed by those who read Arendt as phallocentric and as a "masculine" woman), feminist philosophers now take concerns arising from feminist philosophy as points of departure, as they engage either critically or positively with this tradition, rather than leaving existing forms of thought untouched (Grosz 1990, 60–61).

For Arendt, the central political questions of her time were not those connected to being a woman; it was as a Jew, a refugee, a stateless person that history became a personal life force. These historical facts and contexts provided the frame for her theoretical reflections on racism, totalitarianism, and genocide, on the right to have rights and the threat to these rights posed by statelessness. As philosophers reading her work today, including those of us who are feminist philosophers, it is her reflections on political life—its conditions and structures, its normative claims and risks—that provide points for reflective engagement.

Here, like some earlier feminist philosophers, I carry forward Arendt's concept of natality—but not reading it as an opening to express mothering or women's life experiences. Instead I unpack its methodological and normative implications, and pursue the implications of natality for reflecting on the current refugee crisis.

10.2 Natality as a Concept of the Political

Arendt's concept of natality arguably makes a revolution in philosophical thought, with its inherited focus on mortality as the horizon on which meaning takes place (Schott 2010, 52).[5] Arendt writes of natality under three aspects: as a natural category, as part of the human condition, and as a political category. As a natural category, Arendt used the term birth to refer to the coming into being of individual life and the maintenance of the life of the species. In *The Human Condition* she suggested that this sphere is strictly pre-political, and she lamented the way in which under modernity "the life process itself… has been channelled into the public realm" (Arendt

[5]The following four paragraphs are drawn from Schott (2010, 54–5).

1958, 45). As a feature of the human condition, natality signifies the principle of beginning—the beginning of something unique—in the world (Arendt 1958, 177–8). As a political category, natality refers to the fundamental conditions of life in the political community. It is the loss of natality that marks the loss of genuinely political life under totalitarianism and in the death camps.

For Arendt, biological life and the sphere of necessity belong to the private sphere of the household, whose purpose is the maintenance of life. In her view, force and violence are involved in order to master necessity. But since violence cannot be justified in the political sphere, Arendt viewed biological life as pre-political. In the private sphere, human beings exist "only as a specimen of the animal species man-kind" (Arendt 1958, 46). By implication, the natural life of the body is private and does not display features of the political, including communication which is central for political life. For example, Arendt describes extreme bodily pain as the most private and least communicable of experiences (Arendt 1958, 50).[6] The public sphere, by contrast, is not bound by nature. In order to establish a public sphere of relations between equals, it is necessary that the body's natural needs are satisfactorily met in the household so that they do not impinge on political life. In view of this dichotomy between private and public spheres, some feminist philosophers map a gender dichotomy onto Arendt's text, arguing that Arendt fulfills a "phallocentric desire for release from the realm which 'Woman' has been traditionally figured: bodily maintenance, necessity, and life" (Brown 1988, 180, Dietz 1995, 26).

Arendt introduces the term natality when writing about the most general conditions of human existence: "birth and death, natality and mortality" (Arendt 1958, 8). The most general conditions of human existence also include "life itself…worldliness, plurality, and the earth" (Arendt 1958, 11). Her notion of the human condition emphasizes the fact that human beings are *conditioned*, and that they act out of these conditions in order to create new conditions by either accepting or rejecting this facticity. This notion of givenness, which Arendt develops in her book *Love and Saint Augustine*, is central for her analysis of natality, for natality implies that human beings are created and not just creators. Arendt views natality as the "principle of beginning" (Arendt 1958, 177) in the world, in which the three forms of activity—labor, work, and action—are rooted.

For Arendt, action is the human initiative of beginning something new and is the political activity par excellence. Natality is the quality in the human condition that makes possible human action. She writes, "the new beginning inherent in birth can make itself felt in the world only because the newcomer possesses the capacity of beginning something anew, that is, of acting. In this sense of initiative, an element of action, and therefore of natality, is inherent in all human activities. Moreover, since action is the political activity par excellence, natality, and not mortality, may be the central category of political, as distinguished from metaphysical, thought" (Arendt 1958, 9). She later describes both action and speech as a kind of "second birth, in

[6]The difficulty in communicating extreme pain is the subject of Elaine Scarry's book, *The Body in Pain* Scarry 1985).

which we confirm and take upon ourselves the naked fact of our original physical appearance" (Arendt 1958, 176–7).[7]

There are numerous questions and objections to the way in which Arendt defines the political, and Seyla Benhabib warns against taking the "categorical oversimplifications" in the opening pages of *The Human Condition* as reflective of Arendt's full and complex view of public, political life (Benhabib 1996, 139). Moreover, Benhabib emphasizes the "*normative core of the Arendtian conception of the political*: the creation of a common world through the capacity to make and keep promises among a plurality of humans who mutually respect one another" (Benhabib 1996, 166). Not only can one question whether Arendt can adequately justify these normative comments, but one can also interrogate her sharp distinction between the political world and the economic, the social and the intimate spheres, as the economic realm and even household labor are permeated by political relations (Benhabib 1996, 158).[8]

Here I will examine the connection between natality and Arendt's methodological and normative commitments, and argue that the concept of natality is productive for political reflections on the worst refugee crisis since WWII.

10.3 Natality and the Unprecedented

Natality calls attention to the historical nature of political life and events, in particular to the newness and possibility of disruption of the usual way of being. In this sense, natality is an important concept for Arendt's reflections on revolution, in which she explicitly highlights the language of beginnings. In *On Revolution*, she wrote, "revolutions are the only political events which confront us directly and inevitably with the problem of beginning". In Arendt's view, the modern concept of revolution includes both the concept of history as a new beginning, and the normative claim that the new beginning should coincide with the idea of freedom (Arendt 1963, 21, 28, 29). About modern revolutions she notes, "although there were enough words in premodern political language to describe the uprising of subjects against a ruler, there was none which would describe a change so radical that the subjects became rulers themselves" (Arendt 1963, 41).

---

[7] Arendt began to address the notion of beginnings in her doctoral dissertation, *Love and Saint Augustine*, which was published in German in 1929. She introduced the term natality in *The Human Condition*, published in 1958. When she set about revising her dissertation for English publication in the 1960s, she inserted the term natality in the earlier text to clarify her discussion of beginnings (Arendt 1996). In the first translation of this passage from her dissertation Arendt writes, "Because the world, and thus any created thing, must originate, its being is determined by its origin (*fieri*)—it becomes, it has a beginning." In the revised version Arendt writes instead, "To put it differently, the decisive fact determining man as a conscious, remembering being is birth or 'natality', that is, the fact that we have entered the world through birth (Arendt 1996, 132–3)."

[8] However, below I will argue that this distinction can be very important in reference to the protection of political rights of refugees.

But beginnings cannot just be identified with successful revolutions that support the principle of freedom, which is the frame in which Arendt understood the American Revolution. Beginnings also mark the beginnings of epochs that radically alter the terrain of human institutions and indeed human nature itself for the worse. When Arendt developed her analysis of totalitarianism and the Nazi concentration camps, she draws on the concept of the unprecedented. She compares concentration-camp society to medieval pictures of Hell, noting that the former lacks the hope of justice and grace in the Last Judgement that was present in the latter. She describes the mass manufacture of corpses as preceded by the manufacture of *living* corpses and notes, "the silent consent to such unprecedented conditions are the products of those events which in a period of political disintegration suddenly and unexpectedly made hundreds of thousands of human beings homeless, stateless, outlawed and unwanted, while millions of human beings were made economically superfluous and socially burdensome by unemployment" (Arendt 1951/1973, 447).

The concept of the unprecedented refers to something that is unique, singular, and is not a logical outgrowth of pre-existing patterns or structures. If something has a precedent, it means that it is recognizable in terms of earlier tendencies. But if a political formation is unprecedented, it cannot be understood through the model of continuity or through a model of functionalism. But it was just such interpretations of Nazism that proliferated among sociologists of the period. Sociologists interpreted National Socialism either as a continuation of earlier patterns of bureaucracy or of charismatic leaders, or as a function of a deeper substratum of ideology or of anti-Semitism. But such an approach to a historical event like the Nazi death camps assimilates it to conventional categories (Baehr 2002, 809). Instead, Arendt argued that there were no tendencies that could have led one to predict the emergence and character of a Nazi state (Baehr 2002, 823). Unprecedented events signal "a realignment of the social and political order per se. They create new modes of normality and familiarity, including the familiarity of horror" (Baehr, 825). To try to explain such events in terms of received categories ends up normalizing the phenomenon (Baehr 2002, 826). Hence, new categories of thought are required to capture these changes in the social and political order.

But drawing on the notion of the unprecedented as a methodological approach is not without difficulties. How do we recognize new political formations, as Arendt claimed she did with the emergence of totalitarianism? How can we identify events that are unprecedented, if such a judgment requires that we differentiate them from preceding events and judge them to be utterly dissimilar from previous political formations (Baehr 2002, 822)? Unprecedented entails that events differ sharply from anything previously known, institutionally embedded, or habitually anticipated: "The outcome is something not simply new but sui generis and *original*" (Baehr 2002, 823).

As actions are temporal processes, one must ask when the unprecedented begins and when it ends, since as a process it has no clear time limits (Baehr 2002, 824). Is such a distinction possible at all? Moreover, as the unprecedented sets a precedent for future actions, it becomes absorbed into new modes of normality and familiarity.

At what moment can its unprecedented character be revealed before it becomes part of everydayness?

Hence, the unprecedented is not a method that can be applied instrumentally to any new event, but it is always embedded in historical contexts to negotiate this difficult terrain of understanding. But the concept of the unprecedented requires us to make distinctions, to understand process, and be aware of how changes become sedimented in political life, and as such it is an interpretive reflection of natality. As the unprecedented connotes a movement away from origins, it might appear to function as the opposite of natality. But such an impression wrongly aligns natality with a fixed point of origin, instead of seeing it as a process of renewal and transformation.[9]

Insofar as Arendt treats the unprecedented as a methodological concept rooted in her understanding of natality as the political activity *par excellence*, it does not have normative implications. While she invokes the term unprecedented in her analysis of the French and American revolutions, she also invokes it in her analysis of the radical annihilation of freedom under totalitarianism and in the concentration camps.

10.4 Natality and Ethics

However, natality emerges as a category with ethical content in judging political phenomena which appear as unprecedented. It is the quashing out of human spontaneity that characterizes the concentration camps. In *The Origins of Totalitarianism*, she writes, "The camps are meant not only to exterminate people and degrade human beings, but also serve the ghastly experiment of eliminating, under scientifically controlled conditions, spontaneity itself as an expression of human behavior…." (Arendt 1951/1973, 438). It is this ability to create an isolated world of total domination that made the camps, in Arendt's view, the central institution of totalitarian power. Instead of human spontaneity, the camps created "superfluous human material" (Arendt 1951/1973, 443). And it is in this context that she introduces the term radical evil "previously unknown to us…." (ibid.).

The radical evil of the camps refers to the destruction of what is central for the human condition, the destruction of the possibility of spontaneous action both on the part of inmates and on the part of guards. As such, it implies the "inexorable doom for human beings" (ibid.).[10] As unprecedented phenomena require unprecedented concepts, Arendt invokes the term *radical* evil, well knowing that Immanuel Kant coined this concept before her. Kant, in Arendt's view, tamed the concept of radical

[9]In a similar vein, Foucault distinguishes Nietzsche's genealogy from a search for "lofty origins." Rather, genealogy follows descent: "it disturbs what was previously considered immobile; it fragments what was thought unified; it shows the heterogeneity of what was imagined consistent with itself (Foucault 1977, 143, 147)."

[10]Which doom is also implied by the use of the hydrogen bomb, as Arendt points out.

evil by defining it in terms of "perverted ill will" (Arendt 1951/1973, 459),[11] hence rationalizing it by explaining radical evil in terms of comprehensible motives.

Yet in Susan Neiman's interpretation of *Eichmann in Jerusalem*, it is the very *comprehensibility* of evil that is crucial for Arendt (Neiman 2002, 303). It is an insistence that evil is not supernatural, divine, or demonic, that it is not a threat to reason itself. One might understand Arendt's analysis of Eichmann by recalling her earlier work on Augustine. Augustine distinguished "the well-ordered man" from the "evil man" as a part that has become wicked because it did "not agree with its whole" (Arendt Arendt and 1996, 65). The mis-relation between part and whole is central for Arendt's understanding of the inner dialogue with oneself which is the source of thinking, which she attributes to Socrates' notion that I am my own partner in thinking, and my own witness in acting (Arendt 2003, 90). Adolf Eichmann's "sheer thoughtlessness" (Arendt 1963/92, 287) reflects just such a mis-relation within the unity of the self[12] and the unity of humanity. Hence, like Kant, Arendt also viewed radical evil as comprehensible. But the radicality of evil that Kant did not understand was that human beings could actually will evil for its own sake (Arendt 2003, 62).

In these reflections on the unprecedented events of Nazism and on evil, there are two levels on which the concept of natality operates with normative force. First, natality as referring to the human *condition of plurality* has a positive value for Arendt. Plurality exists both on the subjective level of the two-in-one relation of a person with herself/himself; and plurality exists on the political and intersubjective level as the condition for political life. Annihilating plurality by turning human beings into superfluous matter is a violation of the most fundamental conditions for human existence and hence is evil. Second, natality also refers to the *dynamics of plurality* and the immanent possibilities and risks within the self–self and self–world relation. There are luminous examples of humans getting ethical behavior right, also in extremity, as in Arendt's reporting from the Eichmann trial of the story of Anton Schmidt, a German sergeant who aided the Jewish underground, and was subsequently arrested and executed (Arendt 1963/92, 230). But Eichmann provides us with a vivid example of human beings getting it wrong, and of the enormity of the consequences of such actions in terms of crimes against humanity and genocide. Hence, the dynamics of plurality imply that ethics is a relational project. The relational content can affirm the human condition of plurality; but there is also an ethical risk that a mis-relation, or moral failure (Schott 2003a, b, 93–95) will undermine this condition.

As a method (in relation to the notion of the unprecedented), natality does not have normative content. But natality orients us both to the conditions of human

[11] In "Religion within the Boundaries of Mere Reason" Kant writes, "the human being is *evil* cannot mean anything else than that he is conscious of the moral law and yet has incorporated into his maxim the (occasional) deviation from it." (Kant 1793/2001, 9). And, "This evil is *radical*, since it corrupts the grounds of all maxims…." (Kant 1793/2001, 83).

[12] Arendt reminds us that if one murders, then one has to live with a murderer for the rest of one's life. This, however, might be more manageable than she fathomed. See, for example, Joshua Oppenheimer's film, "the Act of Killing" on the mass murders in Indonesia in the 1960s, in which he asks perpetrators to re-stage their crimes.

existence, and to the dynamics of plurality that structure the relations of ethical life. This multiplicity of meanings may raise general concerns about Arendt's approach. Is natality too general a concept to sustain a consistent meaning? Are concepts that are introduced to clarify fundamental human conditions subject to so many varied mediations that their explanatory power is necessarily limited? Does the tension between normative and non-normative meanings of natality show the inherently ambivalent nature of the concept, with a logic of instability (Schott 2003b, 241)?

Despite these potential disturbances between methodological and normative meanings, I argue that natality does retain an ethical and political force in addressing one of the central challenges of our time, the current refugee crisis.

10.5 Natality and the Present Refugee Crisis

The uptake of Arendt's notion of natality within sociopolitical philosophy has been rather limited, compared to other conceptual agenda-setters such as Michel Foucault's notion of biopower or Orlando Patterson's and Zygmunt Bauman's notion of social death. A quick Google search showed that while natality gets 344,000 hits in 0.32 s, and Arendt and natality get 26,900 hits, biopower reaches 410,000, Foucault and biopolitcs 209,000, and social death 29,300,000.[13] One can recall Giorgio Agamben's comment in *Homo Sacer* about how Foucault summarizes the processes by which "natural life begins to be included in the mechanisms and calculations of State power, and politics turns into *biopolitics*" (Agamben 1995/98, 3). Agamben notes that almost 20 years before Foucault's *History of Sexuality*, Arendt had analyzed the process by which biological life came to occupy the political scene of modernity. He adds, "That Foucault was able to begin his study of biopolitics with no reference to Arendt's work (which remains, even today, practically without continuation) bears witnesses to the difficulties and resistances that thinking had to encounter in this area" (Agamben 1995/98, 4).

From the point of view of the sociology of knowledge, one can explore what factors are operative in the diffusion of concepts. When a concept emerges in a local debate (such as with the emergence of the concept of bullying), it may be taken up by local or national media and become linked to perceptions of the leading social problems of the day. As a term enters public discussions with an array of participants, the term in effect has multiple authors. In the process, the concept becomes entwined with several other recognized social problems, which allows it to gain credibility in the process of "issue-symbiosis" (Schott and Søndergaard 2014, 4–5). Through the process of conceptual contagion, concepts absorb other current issues. Furthermore, different organizations or actors emphasize those parts of the concept that are relevant to them, feeding these examples back into public debates in the media, literature, theatre, and so on.

[13]Google search on July 6, 2016.

Foucault's notions of biopower and biopolitics, which addresses the truth discourses about the vital character of living human beings, have had broad uptake in debates about life, health, modes of subjectification, governance, and security. Paul Rabinow and Nikolas Rose, for example, import the notion of biopower into their analysis of developments about race, population and reproduction, and genomic medicine (Rabinow and Rose 2006, 195). Meanwhile, Arendt's notion of natality has a narrower reach, appearing primarily in discussions of her reflections on politics, mimesis—as in discussions of whether the temporality of natality implies merely repetition of the past or the inauguration of the new (Birmingham 2006, 64)—and education. But it also appears in discussions of war rape and genocide (Schott 2011) and of refugees and human rights (Benhabib 2004; Larking 2014).

Arendt's notion of natality has a great deal to offer for grappling with the present refugee crisis. We are witnessing the greatest number of refugees since WWII, caused by the many conflicts in Iraq, Syria, Eritrea, as well as by the situation in Libya, a transit land riven by war, instability and that has become a haven for smugglers of human beings. The Mediterranean has become a watery graveyard not only for refugees,[14] but also a graveyard for European humanism and universalism (Lucht 2014).

In part, political reflection involves debates about what policies are adequate to address the refugee crises. One can point to the inadequacy of the new EU 10-point plan that prioritizes stopping smugglers rather than offering options for legal application for migration from home countries, or addressing the problems of poverty and hopelessness that lead desperate people to risk their lives at sea (Kleist and Sørensen 2015).[15] But responses to concrete political policies are oriented by an understanding of human rights, freedom, and the conditions for political communities. Arendt's analysis of superfluousness is particularly apt to clarify the stakes in this massive crisis. The Nazi concentration camps provided Arendt an example of how human beings were turned into superfluous human material. This was, in Arendt's view, nothing less than a radical experiment with human nature itself, and the effort to radically transform human beings and eliminate spontaneity and freedom should be considered evil. In the current situation, refugee camps have become part of a permanent state of exception, in which the normal order has been suspended (Agamben 1995/98, 169–70), in which refugees may live for years in camps in which children do not have access to education, and in which girls and women do not have access to basic sexual and reproductive health care. The weekly death toll of refugees drowning in the Mediterranean appears as another manifestation of a willingness among many Europeans to treat human beings as a mass of superfluous matter. And when refugees do arrive across the borders of Europe, they are met by an abdication of

[14] 1350 migrants drowned during the course of just one week in April 2015.

[15] If one established local areas from which people could seek asylum, it would also help correct for the gender imbalance that emerges when young, single men are more apt to take the dangerous journey.

responsibility by European governments.[16] In an Arendtian key, these responses can also be considered evil.

One might object to introducing the term evil into this debate, as it has been widely abused in political rhetoric, as with Ronald Reagan's Star Wars rhetoric denouncing the Soviet Union as the "Evil Empire", or George W. Bush targeting Iran, Iraq, and North Koreas as "the axis of evil". But it is important to distinguish between philosophical concepts and the political rhetoric of evil. The political rhetoric does not mean that evil has no meaning in moral, social, or political domains (Schott 2007, 3–4). Policies which do not effectively stop death at sea, as well as policies which prevent refugees participating in the life of the host country, produce conditions of superfluity. When the Danish parliament in January 2016 approved legislation to require those who have been granted asylum to wait 3 years for family reunification, they were tearing at the remaining shreds of a connection to family for people who have been displaced from their homes. Following the sexual assaults in Köln on New Year's Eve 2016 by illegal immigrants from Northern Africa, a wave of anti-refugee sentiment in Denmark, Sweden, and Germany attests to a mood and mode of nationalism which identifies the rights of residence with blood, culture, and social mores, instead of recognizing that the political sphere is distinct from social, religious, or cultural values. Although Arendt has been widely criticized for insisting on the distinction between the social and the political, it provides a bulwark for recognizing political rights even when social mores of newcomers differ from those of the dominant culture. Recognizing the political condition of natality obligates nations individually and collectively to ensure refugees' rights of family belonging, protection, and participation in the political life of the host country. Differences and contestations over social and religious norms should be recognized and public debates and protests should take place, when relevant. But if refugees are excluded because of their "foreignness", we witness a development that can spiral into excluding the voices of citizens themselves who become categorized as foreigners, as happened with the internment of Japanese-Americans in camps in the U.S. during World War II (Schott 2009, 192).[17]

Hence, there is a deep resonance of Arendt's concepts of natality, superfluous human material, and evil with the refugee crisis facing Europe. I have argued that natality has both a methodological and a normative component. Can one consider the current refugee crisis to be unprecedented? In many respects, it is not unprecedented. It is the consequence of long-standing civil war, insecurity, and crimes against humanity in Syria; of the re-emergence of great power polarities between Russia and the U.S.; of global inequalities between the North and South. And yet the present crisis does pose fundamental challenges to the European political project. The EU scheme to redistribute refugees has by January 2016 only made 4,000 places available, with

[16]Larking suggests this is a sad echo of the abdication of responsibility of governments in Western Europe during the inter-war period. It was the police who took over responsibility for refugees, and they were given license to act with impunity (Larking 2014. 133).

[17]In January 2016, during a rampage against foreigners in the train station in Stockholm, a Swedish teenager was attacked because he had dark hair.

only 272 actually being relocated. European Commission President Jean-Claude Juncker said at this rate, the program "will take until 2101" (cited in Tassinari 2016) to relocate refugees. For Europe to allow (semi-) permanent camps, mass drownings, and a hollowing out of rights guaranteed by the Convention relating to the Status of Refugees signifies a re-alignment of the social and political order.

Thus, the concept of natality focuses attention on the need to protect the political rights of refugees and to challenge the current trend to identify the political sphere with cultural, ethnic, racial, or religious identities. Moreover, it is a reminder that it is as individuals that refugees are subjects of justice, law, and rights (Larking 2014, 135), and they should not be treated as the scum of the earth. Only in this way is it possible to have not only political, but also ethical relations with these newcomers, and to pay respect to the European commitment "Never Again" to allow such evil to take place on its continent.[18]

10.6 Concluding Reflections

As I indicated in my opening remarks, Arendt herself had little interest in the question of women in the philosophical canon. But her concept of natality and its links to reflections on difference, contestation, and multiplicity have struck a chord with many contemporary (and feminist) philosophers. Her philosophical import is not connected with the question of women or of gender in philosophy. Rather, it is through her historical and theoretical examinations of the conditions for human life, for political life, for ethical life—and for their transgressions—that she has become a powerful intellectual figure. In focusing on the necessity of plurality, newness, and spontaneity for political life—and the tragedy of the production of human superfluousness—Arendt offers distinctions and priorities that provide a necessary orientation for the present. We must defend a conception of an open political sphere that is constantly renewed by newcomers, and that is not reducible to social or cultural identities, in order to protect refugees as obligated by both international conventions and moral relations. Without these fundamental protections, we not only risk violating the rights of refugees who seek protection in Europe, but we also risk entering into a process in which citizens risk becoming foreigners in their own country. This, alas, would not be unprecedented.

In addressing such questions, Arendt considered herself to be an engaged political theorist, not a philosopher. In the 1964 interview with Günter Gaus, she argued that she had said goodbye to philosophy once and for all. She did not feel like a philosopher, she was not accepted in the circle of philosophers, and she objected to the claim that she wrote as a political philosopher. There is a difference, she argued, in the nature of the material between philosophy and political theory. In reflecting on issues

[18]Since the civil wars in former Yugoslavia in the 1990s, it has been widely recognized that the imperative "Never Again" has not had the force to stop crimes against humanity and genocide. It may, however, retain force as a regulative ideal to guide moral and political experience.

of ontology and nature, philosophers speak neutrally and objectively for all mankind. But within politics, and hence within reflections on the political, there is always a tension between thought and action. Hence, one cannot be neutral with regards to politics, and she distances herself from the traditional philosophical contempt for politics.

One might be tempted to argue that Arendt changed the conception of philosophy to include this enlarged domain of engaged thinking. She would have insisted that this is not the case. But as the practice of philosophy has changed over the last 50 years in the wake of theoretical contributions from people with social histories of disempowerment within philosophy (Card 1996, 1), many philosophers have challenged the view that philosophers speak in a universal and neutral voice. Insofar as the field of philosophy has changed, one could drag Arendt kicking and screaming in protest back into it. In this sense, the expanded circle of philosophy has also contributed to a changed reception of her work. As with Simone de Beauvoir, Arendt's theoretical practice has made a decisive mark on the changing character of philosophy. One lesson from Arendt's work is that we must reflect as engaged thinkers on the political crises of our times, of which the current refugee crisis is central. If the traditions and circles of philosophy want no part of this type of political reflection, then perhaps we too should say goodbye to these philosophical circles once and for all.

References

Agamben, G. (1995/98). *Homo sacer; sovereign power and bare life* (D. Heller-Roazen, Trans.). Stanford: Stanford University Press.

Arendt, H. (1951/1973). *The origins of totalitarianism*. New York: Harcourt Brace Jovanovich.

Arendt, H. (1957/1997). *Rahel Varnhagen; The life of a Jewess* (L. Weissberg, Ed.; Richard & C. Winston, Trans.). Baltimore: John Hopkins Press.

Arendt, H. (1958). *The human condition*. Chicago: University of Chicago Press.

Arendt, H. (1963/92). *Eichmann in Jerusalem; A report on the banality of evil*. New York and London: Penguin.

Arendt, H. (1963b). *On revolution*. London: Penguin.

Arendt, H. (1996). *Love and Saint Augustine* (J. V. Scott & J. C. Stark, Ed.). Chicago: University of Chicago Press.

Arendt, H. (2003). *Responsibility and judgment* (J. Kohn, Ed.). New York: Schocken.

Baehr, Peter. (2002). Identifying the unprecedented: Hannah Arendt, totalitarianism, and the critique of sociology. *American Sociological Review, 67*(6), 804–831.

Beauvoir, S. de. (192/1974). *The second sex*. New York: Vintage Books.

Benhabib, S. (1996). *The reluctant modernism of Hannah Arendt*. London: Sage.

Benhabib, S. (2004). *The rights of others; aliens, residents and citizens*. Cambridge: Cambridge University Press.

Birmingham, P. (2006). *Hannah Arendt and human rights*. Bloomington: Indiana University Press.

Brown, Wendy. (1988). *Manhood and politics*. Lanham: Rowman and Littlefield.

Card, C. (1996). *The unnatural lottery; character and moral luck*. Philadelphia: Temple University Press.

Dietz, M. G. (1991). Hannah Arendt and feminist politics. In C. Pateman & M. L. Shanley (Eds.), *Feminist interpretations of political theory* (pp. 232–252). Oxford: Polity Press in Association with Basil Blackwell.
Dietz, M. G. (1995). Feminist receptions of Hannah Arendt. In B. Honig (Ed.), *Feminist interpretations of Hannah Arendt* (pp. 17–50). University Park: Pennsylvania State University Press.
Foucault, M. (1977). *Language, counter-memory, practice; selected essays and interviews* (D. F. Bouchard & S. Simon, Trans.). Ithaca: Cornell University Press.
Grosz, E. (1990). Contemporary theories of power and objectivity. In S. Gunew (Ed.), *Feminist knowledge; critique and construct* (pp. 59–120). London: Routledge.
Honig, B. (1995a). Introduction: The Arendt question in feminism. In B. Honig (Ed.), *Feminist interpretations of Hannah Arendt* (pp. 1–16). University Park: Pennsylvania State University Press.
Honig, B. (1995b). Toward an agonistic feminism: Hannah Arendt and the politics of identity. In B. Honig (Ed.), *Feminist interpretations of Hannah Arendt* (pp. 135–166). University Park: Pennsylvania State University Press.
Kant, I. (1793/2001). *Religion and rational theology* (A. W. Wood & G. Di Giovanni, Trans., Ed.). Cambridge: Cambridge University Press.
Kleist, N., & Sørensen, N. N. (2015, April). EU's Løsning på Europas Flygtningskrise er Utilstrækkelig. DIIS impact. Retrieved February 1, 2016, from http://pure.diis.dk/ws/files/118718/diis_impact_drukneulykker_paa_middelhavet_WEB.pdf.
Larking, E. (2014). *Refugees and the myth of human rights; life outside the pale of law*. Farnham and Burlington: Ashgate.
Lucht, H. (2014, December 23). Cemetery in the Sea betrays Europe's moral crisis. Retrieved June 12, 2015, from www.huffingtonpost.com.
Neiman, S. (2002). *Evil in modern thought; an alternative history of philosophy*. Princeton: Princeton University Press.
Rabinow, P., & Rose, Nikolas. (2006). Biopower today. *BioSocieties, 1*, 195–217.
Scarry, E. (1985). *The body in pain; the making and unmaking of the world*. Oxford: Oxford University Press.
Schott, R. M. (2003a). Beauvoir on the ambiguity of evil. In Claudia Card (Ed.), *The Cambridge companion to Simone de Beauvoir*. Cambridge: Cambridge University Press.
Schott, R. M. (2003b). *Discovering feminist philosophy; knowledge, ethics, politics*. Lanham: Rowman and Littlefield.
Schott, R. M. (2007). Evil, terrorism, and gender. In R. M. Schott (Ed.), *Feminist philosophy and the problem of evil*. Bloomington: Indiana University Press.
Schott, R. M. (2009). Kant and Arendt on hospitality. In *Jahrbuch für Recht und Ethik; annual review of law and ethics* (Vol. 17, pp. 183–94). Berlin: Duncker & Humblot.
Schott, R. M. (2010). Natality and destruction. In R. M. Schott (Ed.), *Birth, death, and femininity; philosophies of embodiment* (pp. 49–69). Bloomington: Indiana University Press.
Schott, R. M. (2011). War rape, natality, and genocide. *Journal of Genocide Research, 13*(1–2), 5–21.
Schott, R. M., & Søndergaard, D. M. (Eds.). (2014). *School bullying; new theories in context*. Cambridge: Cambridge University Press.
Tassinari, F. (2016, January 13). All for none, and none for all; life in a broken Europe. In *Foreign affairs*. Retrieved January 29, 2016, from https://www.foreignaffairs.com/articles/western-europe/2016-01-13/all-none-and-none-all.
Young-Bruehl, E. (1982). *Hannah Arendt; for love of the world*. New Haven and London: Yale University Press.

Chapter 11
The Feminine Voice in Philosophy

Naoko Saito

Abstract Despite the call for gender equality, philosophy remains male-centered with women somewhat peripheral to the enterprise; this is not only in terms of its physical and visible aspects but also in the way philosophy is done. When it comes to the question of *how* the female voice is perceived, expressed and represented, there still remains an imbalance. The questions of what 'the female voice' might mean and who it is that represents still remains open. In response, this paper proposes to re-place the subject of philosophy and the subject of a woman through an alternative idea of *the feminine voice in philosophy*. By examining Paul Standish's idea of the 'feminine-receptive mode' of language and thought and Stanley Cavell's *feminism of the father tongue*, it tries to reconfigure the female voice without negating its fated biological origin and traits, yet avoid the confining of thought to the constraints of gender divides. It does this in order to represent the *feminine* in terms of an archetype, as an aspect of the very nature of language. The educational implication of this is the conversation of justice as a way of cultivating the feminine voice in philosophy.

11.1 Introduction

In philosophy, there is an increasing movement to make better heard the voices of female scholars. In the Japan Association of Philosophy, there has recently been established a group (constituted both by men and women) to bring more female philosophers into dialogue. In fact, it is recognized in many countries that the number of female researchers in philosophy is small. What it would be for the voice of female philosophers to be properly heard remains a realistic and pressing question.

In response, I will connect this question with the idea of *translation*. Translation is to be understood not merely as the transposing of meanings between different language systems but rather as of more pervasive importance—as the means by which

N. Saito (✉)
Graduate School of Education, Kyoto University, Yoshida-Honmachi, Sakyo-Ku, Kyoto-Shi, Kyoto 606-8501, Japan
e-mail: saito.naoko.6v@kyoto-u.ac.jp

S. Thorgeirsdottir and R. E. Hagengruber (eds.), *Methodological Reflections on Women's Contribution and Influence in the History of Philosophy*, Women in the History of Philosophy and Sciences 3, https://doi.org/10.1007/978-3-030-44421-1_11

human beings undergo ongoing transformation and rebirth, crossing intra- and inter-cultural boundaries. Translation challenges conventional modes of philosophizing, destabilizing the human subject as an autonomous agent. Much as I wish to see the participation of more women in philosophy, my focus is on the ways that the feminine voice destabilizes predominant ways of thinking that harden the dichotomies of reason and emotion, of justice and care, and of male and female. Translation offers an alternative mode of thinking—replacing the subject of philosophy with an orientation toward 'the claim of reason' (Cavell 1979). This refers to the 'subject' in both senses. It refers to the need both to move beyond the detached, neutral, impersonal voice of philosophy, and to re-place philosophy as an academic subject so that it is understood and undertaken in a different way. The double-genitive in Cavell's title indicates that this works both ways. On the one hand, this is an acknowledgement of the demands of reason, demands that must be placed on me. On the other, it affirms the necessity of me offering my voice, expressing my views, saying how I find the world. In this sense reason depends on my making claims: it depends on the pathos of my life. Hence, it is crucial that this work dispels any sense of a dichotomy between reason and emotion.

I shall challenge the assumption that the feminine voice is a subcategory of the female understood in essentialist terms. To confine the idea of the feminine to some kind of an essence of the feminine will ironically end up suppressing the singularity of the feminine voice—its appropriation and 'theft' (Cavell 1994: p. 37). It would be closer to my conception of what is at stake here to think in relation to archetypes, where these are culturally determined but importantly figure aspects of the biological body. Understood in such terms, then, the feminine voice, as I shall try to show, offers a more oblique, yet more radical political criticism.

Two articles by Paul Standish open a way towards a broader sense of translation, and this contributes to the understanding of the feminine as distinct and as becoming. Stanley Cavell's writings on film further illustrate the nature of the singularity of the feminine voice in philosophy. There the finding of her voice by the woman is a matter of 'perfect pitch' (Cavell 1994: p. 39, 47), a phrase that echoes Emerson's idea of the 'Flying Perfect' (Emerson 2000: p. 252)—an idea of perfection without final perfectibility, in the process of translation as ongoing transformation. Let me begin, however, with a concrete example of the fate of the female voice in philosophy drawn from Japan.

11.2 Care Ethics in Translation

Carol Gilligan speaks of the female voice as a 'different voice' (Gilligan 1982/1993), while Nel Noddings talks of a 'feminine approach' (Noddings 1984). Both criticize the dominance of masculine, *logos*-oriented approaches in the human and social sciences. They put a renewed emphasis on the experience of women and the affective dimension of human life. More recently, acknowledging the feminist and aesthetic dimension in her *Political Emotions* (2013), Martha Nussbaum has sought to reclaim the role of emotion in politics and philosophy, and in our lives as a whole: liberal democracy and justice require the cultivation of 'political emotions', especially sympathetic imagination, compassion, and love. Far from being 'just impulses', these involve 'cognitive appraisal' and 'forms of value-laden perception' (p. 6, 17). Emotions are preconditioned by 'normative commitments' (p. 22). What distinguishes Nussbaum's position from Rawls's is her focus on the particularity of human emotion: she works with a 'pessimistic psychology' (p. 9) and the 'perplexities of real human psychology' (p. 10), which she discusses with rich examples. And she seeks to create space in a liberal society for 'dissenting voices' (p. 7).

Emulating such female philosophers, in general, remains a challenging task. But what 'the feminine voice' might mean is still open to question. Before the struggle of woman's rights, there exist a philosophical question, 'what is the significance of *the woman's voice* in philosophy?' Of course, what the feminine voice amounts to is to be understood with reference to the actual voices of women. But is that all? Is the feminine voice to be found only in that way? Through what is the feminine voice sustained?

There is a particular danger of appropriation in cross-cultural contexts. A case in point is that of the translation of care ethics in Japan. 'Care' is a difficult word to translate into Japanese. Certainly, a part of the power of a term such as 'care' is its degree of instability—the word leaving room for multiple interpretations. But such instability takes on further effects in the process of translation. For example, the term 'care' in Noddings' groundbreaking work (1984) reflects her accommodation and appropriation of concepts borrowed from such thinkers as Heidegger, Buber, and Sartre. In Heidegger, the word in question is *Sorge*. When appropriated by Noddings, however, this concept is given more affective, 'feminine' connotations. With Buber, Noddings puts further emphasis on immediate, face-to-face encounter (as her distinction between 'care for' and 'care about' designates). When this originally unstable, multi-faceted concept is translated into Japanese, the translator must choose either '*Kea*' in the *Katakana* language system (where a phonetic syllabary represents foreign words) or an equivalent Japanese term. In the former case, room for further interpretation is kept open, but the fact that the term remains marked as foreign renders it somewhat technical, other than the everyday connotations of the original; in the latter, the language starts to take on further effects. For example, *Kizukai* or *Omoiyari*, words related to care in Japanese (cf. Kawamoto 2005), have heavily sentimental connotations of looking after the weak, as a matter of compassion.[1] Worse, a certain aspect of the original thought can be aggravated in the hands of its language user. In the classroom and practical fields, the language of care is again translated into accessible terms, oftentimes in the style of categorization, in a form easy to be put on the blackboard. The instability of the word in itself allows the

[1] Fabienne Brugère's work on caring has also been translated from French into Japanese (2014). In her original work, the idea of care is connected to the public realm. In her political discourse, the singularity of the female voice and the internal sense of feminine experience is covered over and assimilated into the appealing public discourse, the female voice being heavily associated with the 'weak' and the disadvantaged. Her mode of thinking is originally based on an opposition between reason and emotion.

idea to be easily assimilated into a masculine mode of thinking, according to which either care is related to the female as a matter of biological difference or it is given a technical, masculine twist. Such multiply-layered processes of translation in themselves facilitate the assimilation of the original female voice into a masculine mode of language and thought, whose benevolent paternalism will be little compensation. In the ensuing eulogy to the vulnerability of human existence and the heightened emphasis on our interdependence, the singularity of the female voice and indeed a certain element of toughness in Noddings' original work are covered over. Such suppression of the female voice need not necessarily be conducted by a male philosopher: a female philosopher who relies on the masculine mode of thinking can lend her hand to this too. The point here is to criticize not so much care ethics itself but blindness to the nature of language that can affect forms of suppression. Language is always and already in the process of translation with further room for its unexpected effects beyond the intention of the original user. Translation is not simply a matter of linguistic exchange. It has political effects. To be oblivious to the nature of language is to collude in the suppression. This example from Japan indicates the need to turn from the female understood in essentialist terms to the feminine as an aspect of the very nature of language and voice.[2]

The experience of paradox in the translation of care ethics in Japan also helps to convey something of the refusal of thinking, in dichotomous, categorical terms. Language has its own life: it is always open to the further possibility of translation. Translation is always turned toward the other—though not in the conventional sentimental, self-sacrificial terms that this last phrase too often carries. It is turned towards the other in that its very point is to communicate something to someone, to another; and it must do this by means of an order of meaning other than that of the source language.

11.3 Towards the Re-placing of the Subject of Philosophy

It is true that the feminine voice is *a different* voice. The issue is how to express and convey the very nature of its difference or what is most at stake in this difference and to recognize the transitory and unstable nature of language, and hence its inherent vulnerability to suppression and theft.

It is here that Paul Standish's idea of the 'feminine-receptive mode' of language is of relevance. His book, *Beyond the Self* (1992/2012) contrasts the 'receptive-responsiveness' of the feminine voice with the 'rational-assertiveness' of the masculine. At the center of his criticism lies the idea of the 'limits of language.' This is presented in relation to the other-regarding virtues of *receptivity* and *humility*, effecting a shift of thinking from the 'rational-assertive' to the 'receptive–responsive (feminine)' and a re-placement of the subject of philosophy, a displacement of the

---

[2]My purpose here is not to discuss care ethics in detail. I refer to it rather for illustrative purposes.

excessive prominence of the idea of rational autonomy in modern Western philosophy. The point is not, however, to privatize or mystify the other as the unknown, or to humiliate or abrogate the self, but rather to turn the attention to alternative dimensions of experience and reality.

The other-regarding practice here elucidates not only the suppressed dimension of the feminine voice in philosophy but also the nature of language itself: it necessitates our receptiveness and sensitivity towards what is transitory and obscure, and to the limits of what we can grasp or know.

As other-regarding in this way, the 'receptive-feminine' is underwritten by an idea of translation. The alternative scope of translation (characterized by asymmetry, imprecision, transitoriness) opens a different way of thinking, a way that has been suppressed where the (masculine) oppositional representation of the female has held sway. This alternative route of thinking is something other than a celebration of the 'different voices' of female philosophers.

Standish's two articles 'Social Justice in Translation: Subjectivity, Identity, and Occidentalism' (2011) and 'One World, One Language' (2010) shed light on the very nature of this difference: they help to show why the idea of translation is crucial to our understanding of the feminine as different from the female voice. In the former article, Standish begins by questioning the narrative of *social* justice in today's global context. The prevalence of the term is then connected with the global hegemony of English in research and in practical fields, and the concomitant trend of 'policy borrowing' in forms of cross-cultural exchange where the traffic is predominantly one-way. The circulation and translation of the term 'social justice' in this context—for example, the preference for the imported term *sociaru jasutisu* to the authentic Japanese *shakai seigi*—disclose a narrowly monolingual mentality, which is tantamount to suppression of thought. The argument opens a horizon of thinking beyond Rawlsian conceptions of justice, beyond the sovereignty of the rational subject and beyond the liberal tradition. To illustrate this point Standish refers to Naoki Sakai's *Translation and Subjectivity* (1997), a text that works especially to show how the experience of translation can unsettle familiar ways of thinking, destabilizing settled notions of human identity and criteria of judgment: the very idea of identity comes to seem an import from the West. Hence, while Occidentalism appears to reciprocate with Orientalism, there is, in reality, an important asymmetry here: in the former, the construction of the West by the East is already grafted upon Western notions of identity (which is based upon the epistemological subject), in a kind of 'double grafting' (Standish 2011: p. 77). This shows that apparently *cross*-cultural relationships cannot be simply a matter of exchange with no residuum. Crossing linguistic and cultural borders necessitates the perspective of a broader sense of translation—the experience of human being as linguistic being always with 'the residuum unknown' (Emerson 2000: p. 254), untranslatability, undecidability—that is, with 'no ultimately satisfactory resolution' (Standish 2011: p. 76). Hence, there is the necessity of ongoing revision of the criteria of judgment in relation to shifting and precarious borders. Such experience cannot be reduced to problem-solving: it is rather, in Sakai's words, an experience of 'ab-solution' (Sakai quoted in Standish 2011: p. 76). Drawing a contrast between two candidate terms in Japanese for the human subject (*shutai* and *shukan*)—a critical decision for the early translators as philosophy established itself in Japan, in the late 19th century, Sakai laments the preference that was given to the more epistemological *shukan*. With his emphasis on the experience of translation as *transformative*, by contrast, he prefers the connotations of *shutai*, of the subject as more holistic, and more bodily and culturally involved (Standish 2011: p. 76).

In Standish's 'One Language, One World' again the idea of translation is highlighted in reference to Derrida, and the experience of an abyss in translation is given sharper focus. It is '"an experience of the edge, the edge of the abyss, between two places "that is precisely the space of responsibility and judgment' (Standish 2010: p. 364). This opens a way to 'think between languages' (p. 376). If we couple these thoughts with the suggestions in his earlier book, this points to the thought that the feminine cannot be fully realized in the supposedly consistent and linear terms of an existing vocabulary—whether of the politics of recognition or care ethics. To go beyond the structure of juxtaposition between male and female, reason and emotion, justice and care, an alternative scope of translation is necessary. Obliviousness to the feminine is tantamount to a forgetfulness of the 'abyssal nature of language' (p. 367). Translation here moves beyond oppositional thinking and categorization. This is a resounding of the feminine voice in its originality and singularity.

Standish's argument reveals the political danger: 'ideological position-taking is commonly shored up by monolingualisms of various kinds' (p. 366). Monolingualism in this sense is a blindness to the transitory and abyssal nature of language, and to its limits. From this perspective, the feminine is something untranslatable in human experience—resistant to conversion yet calling for re-sounding. In response to this paradoxical nature of the feminine voice, translation is an oblique way of eroding the ground of the prevalent masculine monolingualism.

There is, as we have seen, the continuing subtle danger of a re-appropriation of the female voice into a masculine discourse—by both male and female philosophers. In the structure of 'double-grafting', a female philosopher can collude with the reproduction of the masculine voice in the identity politics of recognition. In this structure, the singularity of the feminine is covered over. This might be called the theft of the feminine voice in the masculine discourse of monolingualism. The possible theft of the feminine voice shows the necessity for sensitivity to what cannot be said, to the obscure (Saito 2017) and to the peripheral.

To go beyond the structure of opposition between man and woman, reason and emotion, and care and justice, an alternative mode of thinking (as translation) is necessary. The feminine voice is important because it shifts towards a more transitory provisional language. In the next section, a more positive picture of this re-placed subject will be discussed.

11.4 Perfect Pitch and the Feminism of the Father Tongue

At the conclusion of his discussion, Standish points to the American philosophy of Cavell and Emerson (Standish 2011, pp. 77–79), which recognizes translation as an ingredient of our 'most unhandsome' human condition (Cavell 1990) and directs us towards a new way of thinking and language, beyond the monolingualism of Rawlsian discourse of social justice.

To develop this line of thinking, I hope further to shed light on the nature of the feminine voice in philosophy. This is not to negate the significance of female voices. Ideological assumptions regarding the feminine voice (preoccupied with questions of injustice) ironically end up suppressing the singularity of the feminine, even thus muffling the voice of a female philosopher—through appropriation and 'theft' by a masculine voice (Cavell 1994: p. 37).

Let me begin with Cavell's words in the following:

> I propose here to talk about philosophy in connection with something I call the voice, by which I mean to talk at once about the tone of philosophy and about my right to take that tone. (p. 3)

It is not too much to say that Cavell's philosophy is the philosophy of voice. His idea of voice is rooted firmly in the ordinary life of home, the original place of our participation in a language community. Furthermore, he indicates a possibility of conversation and friendship between men and women, a mode of dialogue different from one based upon confrontation and assertion typically seen in political debate.

'Thoreau's claim to philosophy', he writes, 'is said to be a function of his claims of writing, as to awaken the voice.'[3] Cavell's lifelong endeavor within philosophy is to restore a form of 'autobiographical expression' (Cavell 1996: p. 200). Institutionalized philosophy, with its fated arrogance, has brought about a denial of the other—a disparagement or neglect of the common and the ordinary, and, hence, of that distinctively American voice found in Emerson and Thoreau. Cavell's autobiographical self-searching is simultaneously the assiduous process of 'philosophy's self-criticism' (Cavell 2010: p. 500). His act of writing his autobiography, *Little Did I Know* (2010), and the very fact that this book has been published exemplify what has in effect been a lifelong commitment to 'replac[ing] philosophy' (Cavell 1992: p. 130). The interrelated meanings this phrase carries deserve attention. First, philosophy as autobiography is not simply a matter of personal narrative; rather it involves attesting to an Emersonian idea of 'representativeness' (Cavell 1994: p. 11), such that, in the experience of the writer, 'the deeper he dives into his most private, most secret presentiment, to his wonder he finds this is the most acceptable, most public, and universally true (Emerson 2000: pp. 53–54).' This, Cavell wants to say, is fundamental to our relation to language itself:

> From the root of speech, in each utterance of revelation and confrontation, two paths spring: that of the responsibilities of implication; and that of the rights of desire. It will seem to some that the former is the path of philosophy, the latter that of something or other else, perhaps psychoanalysis. In an imperfect world the paths will not reliably coincide, but to show them both open is something I want of philosophy. (Cavell 2005: p. 185)

[3] Stanley Cavell, *This New Yet Unapproachable America: Lectures after Emerson after Wittgenstein* (Albuquerque, NM: Living Batch Press, 1989), 117.

His *Little Did I Know* enacts and illustrates this distinctive way of connecting the private and the public, the personal and the universal, with the hope that the author's own most private voice can eventually be most public.

For Cavell finding voice is a matter of 'translat[ing]' his father's and mother's words, especially in lending his ear to his mother's silent melancholy (Cavell 1994: pp. 38–39). The presence of his mother is echoed in his attention to the voice of the woman in what he calls 'the melodrama of the unknown women' (Cavell 1996). There the moment of leaving, in the 'pain of individuation' (p. 212), is highlighted as the moment of the woman's finding perfect pitch. The singularity of a female voice demonstrated here cannot be assimilated into the discourse of solidarity and sympathy.

In 'the melodrama of the unknown woman', one of the central themes is the 'incorporation, or theft, of the feminine' voice in the name of 'love as possession' (p. 112). Cavell calls this 'mutual victimization, sapping of one another, vampirism'; it is 'what Emerson spots as adultery' (p. 70).

> This condition [of stealing of voice] is the basis and parable of the possibility and necessity in the education of humans, of making language mine, of finding my voice. (Cavell 1994: p. 37).

This is men's 'plagiarizing women's thoughts' as 'a distrust of their own originality'—a form of skepticism (Cavell 2004: p. 111). The stolen voice of a female in the film is not simply a domestic matter: it has the political implications of what Emerson calls 'secret melancholy', or, as Thoreau has it, 'quiet desperation' (p. 97). This is the state in which a democratic, 'political emotion' is obliterated (p. 98). Hence the point of the melodrama is the reclaiming of 'a right to our own desires' (p. 97). In the melodrama of the unknown woman, the woman regains her voice, as exemplified in *Gaslight* by the voice of Paula: 'Now I exist. Now I speak for myself; and in particular, because I speak in hatred and to you, who have always pretended to understand me, and pretended not to understand me, and who I now know will alone understand my every word and gesture' (Cavell 1996: pp. 47–48). The finding of a female voice is a matter of her 'right to words, of her own voice' (p. 57). This is 'having a voice in your own history', in resistance to the 'denial of voice', and to 'a loss of reason, of mind, as such—say of the capacity to count, to make a difference' (p. 58). This is the affirmation to which the Emersonian complaint of 'voicelessness', of 'hyperbolic inexpressiveness' (p. 66), speaks.

It is true that the situation of women is important in Cavell's film theory, but his concern is not primarily with the status of the woman who has been suppressed: it is rather a 'feminine philosophical demand' (p. 221). The point is the claim of reason, and it is a resistance to that philosophy that suppresses the peripheral and the invisible, and that denies anything that does not fall within its grasp. Reclaiming the feminine voice in philosophy calls for 'serious conversation between women and men' (ibid.).

Consider Cavell's words in the following:

> Film's enforcement of passiveness, or say victimization, together with its animation of the world, entertains a region not of invitation or fascination primarily to the masculine nor even,

> yet perhaps closer to the feminine, but primarily to the infantile, before the establishment of human gender, that is, before the choices of identification and objectification of masculine and feminine have settled themselves, to the extent that they will be settled. (Cavell 2004: p. 277)

Cavell provocatively transgresses the gender distinction, bringing us back to a kind of infancy, the state in which our philosophical thinking can start again, the state that Emerson calls 'neutrality' (Cavell 1994: p. 39) before the ideological, political debate begins. 'Neutrality' here is different from the category of the neuter: it requires a process of achievement, of finding one's voice. In relation to this, we all stand as 'infants'. It is an alternative sense of objectivity.

To live the Emersonian perfectionist life requires having an ear to hear the tone of a particular voice. This is the tone of the 'Emersonian and the feminine demands for a language of one's own' (Cavell 1996: p. 221). It may not be too much to say that his attempted re-placing of philosophy seeks to reclaim the *feminine* voice of philosophy. But certainly, Cavell's endeavor in re-placing philosophy cannot be reduced to a matter of incorporating more voices of women in philosophy as an academic discipline.

Cavell's autobiographical writings offer a clue to the tone of such a feminine voice. For Cavell, as mentioned above, it is not exclusively related to the mother, but in some way both to his father and to his mother. His life in philosophy was 'directed to discovering the child's voice', a voice 'denied' in the tension between his father and mother (Cavell 1994: p. 38). His mother, a pianist, had 'perfect pitch' (p. 21). Cavell describes her tendency to keep her feelings in their 'secret places' (Cavell 2010: p. 113). He acquired from his father 'a knack for telling stories', a 'tact or pitch' of story-telling (Cavell 1994: p. 20), and the 'philosophy of the concepts of pawnbroking' (Cavell 2010: p. 115). His father and mother admired each other's talent, but they were in 'despair of harmony', manifested in his mother's 'periodic silences' (Cavell 1994: p. 21). Cavell incorporated into himself his father's 'shame of himself', his 'self-contempt for his failures' in the world (p. 21, 31). Through their argument, Cavell acquired his 'conception of philosophy as the achievement of the unpolemical, of the refusal to take sides in metaphysical positions' (p. 22). Such a view of philosophy is rehearsed in *Little Did I Know* where he interprets Wittgenstein's remark, 'Philosophy leaves everything as it is', as *not* meaning that philosophy 'does not, and indeed should not, rock the political boat' (Cavell 2010: p. 478). Cavell's lifelong endeavor to re-place philosophy is inseparable from the clinical path he has been following from childhood to adulthood.

The '*father* tongue' (my italics) paradoxically has a crucial relevance to the nature of the *feminine* voice of philosophy. In *A Pitch of Philosophy* (1994), and immediately after he discusses the issue of finding his own language, between his father and mother, Cavell refers to the following passage from Thoreau's *Walden* [pp. 39–44] (which is quoted also in *The Senses of Walden* [Cavell 1992, p. 15]):

> It is not enough even to be able to speak the language of that nation by which they are written, for there is a memorable interval between the spoken and the written language, the language heard and the language read. The one is commonly transitory, a sound, a tongue, a dialect merely, almost brutish, and we learn it unconsciously, like the brutes, of our mothers. The

> other is the maturity and experience of that; if that is our mother tongue, this is our father tongue, a reserved and select expression, too significant to be heard by the ear, which we must be born again in order to speak (Thoreau 1992: pp. 68–69).

This is a controversial passage. It may cause an immediate reaction from a reader: Isn't this the disparagement of the mother extended by implication to spoken language (call it the native tongue) and associated with 'the brutes'? Cavell's reading of this, however, is provocatively different. The father here is not an authority figure: the mother tongue (speech) and the father tongue (writing) are not in a hierarchical relationship. In 'Emersonian and Thoreauvian perfectionism' (Cavell 2010: p. 479), the father tongue hinges on rebirth (Cavell 1992: p. 16).[4] This is not, however, a once-and-for-all event, but a daily baptism, a daily requirement that our right relation to language exacts. The idea of the father tongue is inseparable from the idea of translation. The father tongue bears witness to the transitory nature of language, to the fact that the "truth is tranlsated" (Thoreau 1992: p. 217; Standish and Saito 2017: p. 2).

The rebirth of language involves a 'discontinuous reconstitution of what has been said, a recounting of the past, autobiographizing, deriving words from yourself' (Cavell 1994: p. 41). Rebirth requires human beings to undergo testing by 'a line of words so matured and experienced' that they divide you 'through the heart' (Cavell 1992: p. 17). This suggests a kind of death by 'happy injuries, ecstasies of exactness' (p. 44). Re-engagement with the father tongue is a way of sustaining the space of what Cavell calls 'the daily, insistent split in the self that being human cannot… escape' (Cavell 2004: p. 5). The relationship between the mother and the father is not reciprocal, nor is it complementary or symmetrical. There is no 'pure' original state in which they are united, in a perfect fit. Rather they symbolize our dual relation to language.

Cavell's attempt to re-place philosophy might better be expressed in an apparently paradoxical phrase, *the feminism of the father tongue*—a philosophy that is feminine. The feminism of the father tongue releases us from an illusion of face-to-face immediacy that is typically associated with the feminine and motherhood. On the one hand, the feminine demand of language, by contrast, is a matter rather of Emersonian and Thoreauvian 'self-empowering' (Cavell 1996: p. 36). On the other, with Cavell's sensitivity to his mother's silence and to his father's sense of shame, the tone of philosophy here is what Standish has called 'receptive-responsiveness'. While the feminine tone of philosophy attempts to go beyond the division between reason and emotion, this is not to weaken reason but rather to make it more rigorous. For Cavell it is a shift towards the kind of thinking that is at 'some as yet unknown distance from what we think of as reasoning' (Cavell 1990: p. 46). In *The Claim of Reason* (1979), he claims that passion and desire are the sources of our 'search for reason'

[4]'A son of man is born of woman; but rebirth, according to our Bible, is the business of the father. So *Walden*'s puns and paradoxes, its fracturing of idiom and twisting of quotation, its drones of fact and flights of impersonation—all are to keep faith at once with the mother and the father, to unite them, and to have the word born in us' (Cavell 1992: p. 16).

(Cavell 1979: p. 20). This is echoed in and echoes Emersonian thinking as 'the passive (patience, passion)' (Cavell 1990: p. xxxvi), thinking as reception (Cavell 1996: p. 221), and 'thinking as the receiving or letting be of something' (Cavell 1990: p. 38, 1992: p. 132). Standish's and Cavell's voices are *feminine* voices of male philosophers, for the conversation of men and women.

Thus, the father tongue has the nature of the feminine voice. It is related to emotion, passion, receptivity, and yet is not a reactionary turn to irrationalism. Furthermore, it calls, within the philosophical discourse, for a replacing of the subject of woman—whose singularity is not contained in the vocabulary of connection, interdependence, and sympathy. This singularity has a force sufficient to destabilize the vocabulary of care ethics because care ethics is relational ethics. Its transitory nature reconfigures the subject, always in translation.

The feminine voice involves us in the realm of the obscure, the peripheral. To be in tune with such precarious borders of our lives, we have to sensitize ourselves to 'perfect pitch', to the sound of philosophy. As a matter of ongoing perfection, the sounding of perfect pitch is crucially related to Cavell's Emersonian idea of *finding as founding*, to an art of sojourning—'living each day everywhere and nowhere, as a task and an event' (Cavell 2005: p. 229). Cavell's Emerson is not the Emerson of the Over-Soul: it resists teleological salvation, contesting tears. The therapy of philosophy, its finding of voice, is conducted in Emerson's experimental spirit. In 'discontinuous encirclings' (Cavell 1990: p. xxxiv; 1992: p. 128). Cavell presents this forward movement, echoing Emerson's idea of the 'Flying Perfect'—this evanescent moment that constantly betrays our grasp, and yet at the same time, awaits our expression. The continuity of one's life is created from within discontinuity, and this is a kind of continuity without 'foolish consistency' (Emerson 2000: p. 138). The self is always to be found, to be attained. The finding of voice is brought about only by the power of one's own words in conviction. We are continually to be redeemed by words. This is the experience of translation as exemplified by Thoreau's idea of 'standing on tiptoe' (Thoreau 1992: p. 71). This is an antifoundationalist thinking on the way, where American philosophy is continually in translation (Saito 2019). In this ongoing perfection, the transitory sense of perfect pitch refuses to be assimilated and contained in the language of monolingualism.

Cavell describes the singularity of the female voice as follows.

> The position of women is neither that of exiles nor of immigrants: unlike the immigrant, the woman's problem is not one of not belonging but one of belonging, only on the wrong terms; unlike the exile, the woman is not between two different cultures but is at odds with the one in which she was born and is roughly in the process of transfiguring into one that does not exist, one as it were still in confinement. (Cavell 1996: p. 213)

When a perfect pitch is found, there is a kind of leaving of home, the subject always in translation: separation involves 'the pain of individuation'. Such a contesting of tears refuses to be fully translated into the language of care ethics. The finding of perfect pitch requires a different kind of conversation between men and women.

11.5 Conversation as Translation: The Re-Sounding of the Feminine Voice

Finding perfect pitch requires the cultivation of aesthetic imagination. In the feminism of the father tongue, conversation is needed—the 'happy possibility' of friendship and mutual education between man and woman (Cavell 2004: p. 277). Rather than excluding male philosophers from dialogue, *conversation as the experience of translation* between men and women is called for. Transgressing ideological gender borders, this is to retain space for what Emerson calls 'a residuum unknown, unanalyzable'—or, in Cavell's words, 'the opacity, or non-transparence, of the present state of our interactions, cooperative or antagonistic' (p. 173). This is a mode of conversation that resists easy oppositions and categorization. Its virtues are 'those of listening, the responsiveness to difference, the willingness for change' (p. 174). For a woman to 'regain [her] tongue' in conversation (Emerson 2000: p. 210), the other as a friend (a man for a woman, a woman for a man in the film), is a crucial resource in finding the measure of the authority of her speech. It is the site for mutual education between men and women, crossing and redefining vague borders, exposing themselves to the destabilizing of existing frameworks, where selves are translated through release of oneself to the invasion of the other. With these implications conversation as translation is a site for 'philosophy as the education of grownups' (Saito and Standish 2012). This will be a more oblique and yet a most radical way of political criticism to alter the 'imbalance' between men and women.

References

Brugère, F. (2014). *The ethics of care*. Translated by Tetsu Harayama and Rieko Yamashita (*L'ethique du <care>* [Paris: Pfu, 2011]). (Japanese).

Cavell, S. (1979). *The claim of reason: Wittgenstein, Skepticism, Morality, and Tragedy*. Oxford: Oxford University Press.

Cavell, S. (1990). *Conditions handsome and unhandsome: The constitution of emersonian perfectionism*. Chicago: The University of Chicago Press.

Cavell, S. (1992). *The senses of Walden*. Chicago: The University of Chicago Press.

Cavell, S. (1994). *A pitch of philosophy: Autobiographical exercises*. Cambridge, MA: Harvard University Press.

Cavell, S. (1996). *Contesting tears: The hollywood melodrama of the unknown woman*. Chicago: The University of Chicago Press.

Cavell, S. (2004). *Cities of words: Pedagogical letters of a register of the moral life*. Cambridge, MA: The Belknap Press of Harvard University Press.

Cavell, S. (2005). *Philosophy the day after tomorrow*. Cambridge, MA: The Belknap Press of Harvard University Press.

Cavell, S. (2010). *Little did I know: Excerpts from memory*. Cambridge, MA: The Belknap Press of Harvard University Press.

Emerson, R.W. (2000). In: B. Atkinson (Ed.), *The essential writings of Ralph Waldo Emerson*. New York: The Modern Library.

Guilligan, C. (1982/1993). *In a different voice: Psychological theory and women's development.* Cambridge, MA: Harvard University Press.

Kawamoto, T. (Ed.). (2005). *Social ethics of care: Bridging medicine, nursing, care and education*. Tokyo: Yuhikaku. (Japanese).
Noddings, N. (1984). *Caring: A feminine approach to ethics and moral education*. Berkeley: University of California Press.
Nussbaum, M. C. (2013). *Political emotions: Why love matters for justice*. Cambridge, MA: Belknap Press of Harvard University.
Saito, N. (2017). Philosophy as translation and realism of the obscure. In P. Standish, & N. Saito (Eds.), *Stanley Cavell and philosophy as translation: The truth is translated* (pp. 11–23). London: Rowman & Littlefield.
Saito, N. (2019). *American philosophy in translation*. London: Rowman & Littlefield.
Saito, N., & Standish, P. (2012). *Stanley Cavell and the Education of Grownups*. New York: Fordham University Press.
Sakai, N. (1997). *Translation and subjectivity: On "Japan" and cultural nationalism*. Minneapolis: University of Minnesota Press.
Standish, P. (1992). *Beyond the self: Wittgenstein, Heidegger and the limits of language*. Avebury: Aldershot, UK (*Beyond the self: Wittgenstein, Heidegger, Levinas and the limits of language*. Translated by Naoko Saito. Tokyo: Hosei University Press, 2012 [Japanese]).
Standish, P. (2011). Social justice in translation: Subjectivity, identity, and occidentalism. *Educational Studies in Japan: International Yearbook, 6,* 69–79.
Standish, P. (2010). One language, one world: The common measure of education. *Philosophy of Education, 2010,* 360–368.
Standish, P., & Saito, N. (Eds.). (2017). *Introduction. In Stanley Cavell and philosophy as translation: The truth is translated* (pp. 1–9). London: Rowman & Littlefield.
Thoreau, H.D. (1992). In W. Rossi (Ed.), Walden and resistance to civil government. New York: W. W. Norton & Company.

Chapter 12
Iris Murdoch on Pure Consciousness and Morality

Nora Hämäläinen

Abstract This chapter provides a reading of Murdoch's discussion of "pure consciousness" in Chap. 7 of *Metaphysics as a Guide to Morals* (MGM, Murdoch 1992), which is one of the central places in her work where we can find sustained attention to this matter. Here she builds a picture of moral consciousness on a critique of nineteenth and twentieth century thought on consciousness and cognition. Through this reading we can see how Murdoch uses a variety of very heterogeneous accounts of consciousness (and the purification of consciousness), to establish one of the pillars of her moral philosophy: the image of moral work as the work of a singular consciousness upon a world which is real and tangible, and yet veiled by the limitations of the self. Attention to this also helps us to place Murdoch in the context of currently growing interest in philosophy as a self-transformative practice.

Iris Murdoch entered Oxford in 1938 and was part of a quite remarkable cohort of woman philosophers, including Elizabeth Anscombe, Philippa Foot, and Mary Midgely. Anscombe's and Foot's work in the 1950s and 1960s form the starting point of both contemporary virtue ethics and moral psychology in the Anglophone setting. Like her colleagues, Murdoch insisted on the necessity of a rich understanding of human personhood, of moral psychology and the virtues—topics that were very much out of fashion in mid-twentieth century philosophy. Murdoch too contributed insightfully to these issues, in her philosophical writings as well as in her novels. But, where Foot's and Anscombe's influence has to a certain extent been domesticated into a broadly analytic tradition of normative moral theory, Murdoch is today most prominently and most insightfully approached by philosophers who eschew moral theory (Mulhall 1997, Diamond 1996, Gaita 2003, Forsberg 2013). Mainstream analytic philosophers find her difficult to understand.

This positioning of Murdoch as a relative outsider in mid- to late twentieth century Anglophone moral philosophy is something of late construction. She maintained

N. Hämäläinen (✉)
Centre for Ethics as Study in Human Value, Department of Philosophy, University of Pardubice, Stavařov 97, 532 10 Pardubice, Czech Republic
e-mail: nora.hamalainen@upce.cz

S. Thorgeirsdottir and R. E. Hagengruber (eds.), *Methodological Reflections on Women's Contribution and Influence in the History of Philosophy*, Women in the History of Philosophy and Sciences 3, https://doi.org/10.1007/978-3-030-44421-1_12

close intellectual friendships with both Anscombe and Foot. She held a fellowship at St. Anne's College, Oxford from 1948–1963. Although she eventually left Oxford to be a fulltime novelist, she continued to write philosophical texts as well. Her book *The Sovereignty of Good*, first published in 1970, was widely read, and prominent philosophers somewhat younger than her, like Bernard Williams, Charles Taylor, and John McDowell refer sympathetically to her work. And the texts she wrote were very much characterized by a kind of internal critique of the philosophy of her time and place.

On the other hand, she did use her relative freedom to write in a manner quite unlike her analytic contemporaries. She insisted even in the 1980s that "Philosophy aims to clarify and to explain, it states and attempts to solve very difficult highly technical problems and the writing must be subservient to this aim" (1997, p. 4). But her own writing does not speak so much of technical problems as it does of the attempt to articulate an original way of looking.

Murdoch's moral philosophy has sometimes been casually described as a kind of "moral phenomenology". This description is helpful if it is understood in the broad sense of a "what it is like-ness" of morality, rather than as suggesting a connection to the phenomenological tradition. One of the first things that students and scholars learn about Iris Murdoch is that she proposed an ethics of attention, where the individual person and his relation to goodness take center stage. This was to be contrasted with the mainstream of mid- and late twentieth century moral philosophy (Ayer, Hare), where the consciousness of the individual person had been unduly neglected or considered with suspicion.

Her polemic positioning in mid- and late twentieth century Anglophone moral philosophy is well established in commentaries on her philosophical work (Antonaccio 2000; Widdows 2005; Broackes 2011). Less attention has been paid to the details of Murdoch's own substantial ideas on the topic of consciousness. In this paper, I provide a reading of Murdoch's discussion of "pure consciousness" in Chap. 7 of *Metaphysics as a Guide to Morals* (MGM, Murdoch 1992), which is one of the central places in her work where we can find sustained attention to this matter. Here she builds a picture of moral consciousness on a critique of nineteenth and twentieth century thought on consciousness and cognition. Through this reading I investigate how Murdoch uses a variety of very heterogeneous accounts of consciousness (and ideas of purification of consciousness), to establish one of the pillars of her moral philosophy: the image of moral work as the work of a singular consciousness upon a world which is real and tangible, and yet veiled by the limitations of the self.

This chapter of MGM is interesting in its own right, for a better understanding of Murdoch's conception of consciousness and its place in twentieth century ethics. But it is also paradigmatic of Murdoch's philosophical method in MGM and helpful for understanding the book's special kind of systematicity. Thus, I will use it for two distinct but closely interrelated purposes here: (1) To exhibit Murdoch's idea of moral consciousness, and (2) to illuminate Murdoch's idea of philosophical work and the nature of philosophical progress.

Both are usefully illustrated by her favourite imagery of a platonic ascendance from the cave, from the given to a greater truthfulness. Yet, I will also argue that we should not conflate these two strivings: the philosopher's ascent is not, according to Murdoch, identical to the ascent of the moral person, and vice versa. Thus, Murdoch does not represent an idea of "philosophy as wisdom" but retains the modern idea of philosophy as an intellectual pursuit beyond and beside the development of individual wisdom and virtuous personhood.

12.1 Consciousness

Chapter 7 of MGM begins (disconcertingly) with a long quotation from Walter Pater, expressing a celebration of what Murdoch calls "the fundamental reality of the individual" in his act of passionately enjoying his present experience.

> The service of philosophy, of speculative culture, towards the human spirit is to rouse, to startle it into sharp and eager observation. Every moment some form grows perfect in hand or face; some tone on the hills or the sea is choicer than the rest; some mood of passion or insight or intellectual excitement is irresistibly real and attractive for us,—for that moment only. Not the fruit of experience, but experience itself, is the end. (Pater, quoted by Murdoch, MGM, p. 217)

The passage quoted by Murdoch is much longer than this, developing Pater's aestheticized celebration of keen, self-aware experience. This hedonistic/aesthetic spirit is displayed by Murdoch, almost without explicit comment, only to be rapidly contrasted with Simone Weil's both ascetic and ethical "wish that she could 'perceive without reverie'" (ibid. p. 218)—a theme that will recur throughout the chapter. And from there Murdoch moves, without settling the staged debate between Weil and Pater, to emphasize that understanding "the experience of consciousness" is essential for moral philosophy and that it has wrongly been neglected by philosophers. Then she quickly goes through how and why it has been so neglected. Among the obstacles we find the "Kantian positive conception of will" which deprives "ordinary awareness" and "mere feelings" of value and interest; Hegel's and Marx's consciousness which is not that of an individual; the empiricists "contents of consciousness" which seem to be too facile, fragmentary, flat and easily available. She notes that the concept of consciousness "was also denied to moral philosophy, and to philosophy of mind in general, when a rejection of the Cartesian view of consciousness seemed to make any treatment of personal experience seem impossible" (ibid. p. 219). (Here she is thinking of Gilbert Ryle and Wittgenstein).

Murdoch makes the case that the problem of consciousness has been both one of the most central problems of our tradition, and none the less, arguably, neglected, bypassed. The key questions of modern western philosophy are rephrased by Murdoch as follows: "Can I have clear intuitive knowledge of something 'owned' by me, something which is indubitably present to me? Can I 'see' such a thing, can I be said to know it, can this certainty be extended to a public world?" (ibid. p. 220) Descartes and Husserl, she notes, "appeal to consciousness and do so with certain ends in view; the consciousness of consciousness is to reveal the foundations of knowledge" (ibid. p. 238) Murdoch asks if we really believe that such certainty can be had. Certainly, this kind of knowledge is not what *she* is after, but the philosophers' search for it provides is the background against which she stages her own investigation of "the experience of consciousness". Consciousness, Murdoch suggests, has been central for the tradition, but rarely for its own sake and mainly for its role in settling questions about knowledge, truth, certainty, reality, the mind-world issue. The treatment of consciousness has been entirely shaped by other ends in view. What she wants to do is to turn our attention to the contents of consciousness, the experience of consciousness, and how this experience exhibits qualitative differences that are significant for our quest for the good as well as our quest for truth.

There are two different lines of argument present here: one is polemic (and to some extent implicit), attempting to settle the anxieties concerning truth, reality, knowledge that are so central to the western philosophical tradition by showing that they stem from a very specific way of seeing the world, which is far from unavoidable. The other line of argument is substantial—attempting to articulate a specific view of consciousness and the complex work of consciousness, which flouts these western, philosophical anxieties, and opens up for an interest in the quality of consciousness in its own right. I will focus on the latter here, but it is important to keep the former in mind since it forms a constant underlying presence in Murdoch's discussion.

We can briefly contemplate the beginning of the chapter. Pater is quoted to represent an affirmation and celebration of the individual consciousness in its heightened moments of passionate enjoyment, aesthetic experience, or love. What Murdoch scoffs at here is the facile, cheap, romantic, naïve aspect of his way of looking. But she also finds something of special interest here, something that is not annulled through the contrast with Weil's sternness: she takes the idea of a heightened, passionate, intense experience—and then, with Weil, she turns the focus of this experience outwards, towards the object of our attention, emptying it of the romanticized "self". This movement relieves it of self-centred "reverie". The picture evoked is reminiscent of the completely self-forgetful, self-effacing experience of the kestrel, which is one of the central picturesque examples of self-forgetfulness in Murdoch's most widely read book *The Sovereignty of Good*.

> I am looking out of my window in an anxious and resentful state of mind, oblivious of my surroundings, brooding perhaps on some damage done to my prestige. Then suddenly I observe a hovering kestrel. In a moment everything is altered. The brooding self with its hurt vanity has disappeared. There is nothing now but kestrel. And when I return to thinking of the other matter it seems less important. (Murdoch 2001, p. 82)

In MGM, the picture of self-forgetful experience is further opened up and developed.

The purification of consciousness is perhaps not a central theme in the western philosophical tradition (where consciousness has above all been the handmaid of questions of knowledge, truth and certainly), but Murdoch insists that it can be explored through this tradition. She does this by letting loose the Zen author Katsuki Sekida on Husserl. Husserl explored the idea of a pure *cogitatio*, to be achieved through a phenomenological reduction that would let the phenomena appear without prejudice. Wittgenstein, Murdoch notes, would not have had much charity for such an idea of contextless knowledge. Adorno again "regarded Husserl's reduction as a last move of bourgeois idealism in search of a safe world accessible to universal knowledge" (MGM, p. 240). Sekida's objection is different: he takes seriously the idea of a purification of consciousness but questions the idea that something of the kind can be achieved through an act of philosophical reflection. In Zen, the road towards the purification of consciousness is long, arduous and individual, and the self a constant, persistent obstacle.

Of course, Murdoch notes, Zen is not a philosophy but a spiritual tradition, and Husserl and Sekida do not necessarily have the same object in mind. Yet there is something very attractive about this critique for Murdoch. The idea of pure consciousness, to be achieved through a technical philosophical operation, seems to her as confused as it does to Sekida. "Consciousness au fond and ab initio must contain an element of truth-seeking through which it is also evaluated. In this sense, some cognitions are purer than others; but we cannot descend by any unitary 'scientific' or systematic method below the levels at which, in various ways, we test truth and reflect upon moral understanding" (ibid. pp. 241–242). Murdoch's philosophical interest is turned here toward a complex lived world where knowledge is essential but always limited; where the self often stands in the way of proper judgment—an entangled world where knowledge and evaluation are intertwined. Purifying consciousness of distorting interests is for her not a philosophical move to outsmart this overwhelming mess of a world (to attain knowledge), but rather a kind of spiritual work, a slow work of consciousness upon itself, to be performed in this world, as we inhabit it. This is the angle from which she enters her discussion of Zen as a spiritual discipline.

A Japanese poem, quoted by Sekida and Murdoch, should be repeated here:

> Hearing, seeing, touching and knowing are not one and one;
> Mountains and rivers should not be viewed in the mirror.
> The frosty sky, the setting moon–at midnight;
> With whom will the serene waters of the lake reflect the shadows in the cold?
> (Sekida, quoted by Murdoch ibid. p. 242)[1]

"Not be viewed in the mirror", as Sekida explains it, means that the world should not be viewed as a "projection of the subjective mirror of your mind" (Murdoch ibid. pp. 242–243). The self, the projecting subject, is effaced in the poem to bring forth the reality of the "serene waters". In pure cognition "the cognition occurs solemnly and exclusively between you and the object. ...Think of the moment your hand touches the cup: there is only the touch" (Sekida, quoted by Murdoch ibid. p. 243).

What we need is to achieve this access, this presence to the things in the world. Like with Murdoch's kestrel, there is no room here for philosophical preoccupation with the theoretical questions of consciousness and world. Murdoch's question of consciousness is much more concrete and practical: it is plainly about how to be in

[1]This poem is attributed by Murodch to zen master Nansen, but as Fiona Tomkinson observes, in the work quoted, Sekida's bo ok Zen Training, "the poem is actually given as a work by Setcho, the author of the Hekigan Roku (or Blue Cliff Record), commenting on the story 'Nansen Views the Flower'. (Tomkinson 2019, p. 60)

the world, how to look in order to see, how to tune into it and how to educate oneself to greater sensibility. All of this is difficult, but not in a technical way. We have in our ordinary lives these images and moments of "seeing without reverie", without self-preoccupation, but also the recognition that this is difficult and often unattainable. She doesn't say that the philosophers' preoccupation with the possibility of truth, knowledge, certainty, are futile—but she does suggest that consciousness can and should be approached in this quite different way, which is also a way more readily conductive to moral thought as well as to the overcoming of the kind of scepticism that is always close at hand in our philosophical tradition.

In case we are not persuaded by this Zen imagery, Murdoch provides yet another example of the purification of consciousness—that of artistic creation (ibid. p. 246)—quoting Rainer Maria Rilke on the topic of Paul Cézanne. In Rilke's words Cézanne's artistic genius has something "to do with 'animal attentiveness', 'good conscience', 'only doing what you know', 'simple truthfulness', the 'consuming of love in anonymous work'" (ibid. 247)—again descriptions which evoke the idea of a perception ideally undisturbed by self, its hopes, dreams, neuroses and fantasies.

12.2 Consciousness Moralized

We may now want to ask how the above images of consciousness purified relate to Murdoch's view of morality and value as integral parts of our consciousness of the world, and further, how this imagery gets the moral weight it does. In Chap. 7 of MGM Murdoch is not describing the merging of fact and value in consciousness (although that is present too in the idea of disciplined truth-seeking). Yet she repeatedly returns here to the idea that consciousness is moral, that it is shot through with value, etc. (pp. 237, 251, 257–258) Why? A simple way to connect the issue of consciousness as evaluative/moral and the issue of pure consciousness could be the following: the presence of value in consciousness is taken for granted in the picture of consciousness put forward here. This is not the issue that is here under negotiation. The problem or predicament to which the purification of consciousness presents itself as a tentative solution is the situation of a being whose experience is defined by a thorough entanglement of fact and value, knowledge and evaluation, truth and goodness. Our consciousness can be more or less truthful, more or less good. We see the world through our own needs and desires, our cultural and individual biases, hopes and fears, inherited norms, taboos and these distract us from the simple immediacy of things and people around us.

What we need, in Murdoch's view, is not to learn a way of looking that is coloured by the right values or rules of conduct in a given way, something that can be easily passed on in explicit teaching. This is a superficial and partial understanding of what moral education or learning could be. What we need instead is an exercise in attending closely, earnestly—letting the world show itself. This does not imply that Murdoch would assume the possibility of an experience that is not culturally and linguistically mediated. She is highly aware of the limitations of our conceptual resources and also of the fact that they undergo change over time. What she points to is something much less theoretical and quite ordinary, the sense in which we could say: "I didn't *really see* this painting last time I was in Venice; I was in a fog for months after the divorce" or "Paul is so worried about how Lisa judges him, that he can't figure out that, more than anything, she is sad."

In order to see what we have before us more clearly—people, relations, nature, art objects, and situations—we need to practice self-forgetfulness, exercise our awareness of other things. The "right values" or "good principles" will not help us if we are immersed in the fantasies of what Murdoch elsewhere calls our "fat, relentless ego" (2001, p. 51). Zen exercises, prayer, the creation or enjoyment of good art, the learning of new practices like speaking foreign languages or playing music: these are all activities where the relevant kind of self-forgetfulness is naturally exercised, and we can learn by attending to these. What they also teach us is that the purification of consciousness, the pure attention to what is not us, is not a simple once-and-for-all matter. It is theoretically plain and simple, but it is difficult and slow in practice.

Lovibond (2011) has suggested that this emphasis on self-forgetfulness in Murdoch's moral philosophy is politically naïve, if not reactionary, considering the structural nature of the biases in our perception. The central problems of moral consciousness are in Lovibond's view is not selfishness, fears or neuroses, but rather oppressive structures and collective stereotypes that, according to Lovibond, are swept under the carpet by Murdoch's emphasis on the individual. Certainly, there is a need to extend Murdoch's attention to individual biases with awareness of structural biases, but these perspectives are complementary rather than conflicting.[2] Overcoming structural biases is like overcoming individual biases; both involve self-criticism, humility and knowledge of structural features of our present modes of life.

But does value, goodness, enter into the attempt to purify consciousness? How could a purified consciousness of another person tell me how to behave towards him? Couldn't the "purification" lead to a situation where orientation is lost, where we no longer know what is important, what is good, etc? It must be noted that considering the examples Murdoch gives, considering her talk of attention as "loving", the contemplation of a chilly, reductive, behaviourist observation of others is not relevant here. What she urges us to practice is a way of looking at people which takes into account how they each have a life of their own, different from and independent of ours. This is a perspective which also takes seriously the separateness of natural objects and artefacts, and their capacity to speak about things we did not know. Wonder is important here. What she seeks is to evoke a tract of experience, to make it vivid for us, so that we can recognize it in our own lives.

[2] See Hämäläinen (2015) and Clarke (2011).

12.3 Murdoch's Method

All that has been said thus far could, however, strengthen certain misgivings that we may have over Murdoch's method of philosophical writing in MGM. In the chapter discussed she starts with a quotation, which she neither concurs with nor systematically criticizes. Simone Weil makes repeated appearances but is not systematically discussed. The philosophical tradition on consciousness is reviewed but dropped as ultimately uninteresting. Sometimes the writing is messy, even on a very close reading. The pieces on Zen Buddhism, Rilke and Schopenhauer say the essentials. She wants to communicate a certain rather stern and indeed normative perspective on consciousness as central for our moral lives. According to this perspective, if we wish to be good people, responsible moral beings, we must look without reverie, discipline ourselves, and this is slow, hard work. But we may want to ask: couldn't this have been said more concisely?

Yet, the chapter has its own peculiar systematicity: it urges us to consider a certain way of looking at consciousness, then shows how the philosophical tradition's concerns have been different. In a further step, varieties of this way of looking are presented: what it means for a person to pay attention to what is other than herself, what it means to come to see "truthfully". The aim seems to be to invite the reader to adopt a way of looking, to try it on, to try and see what things look like from that perspective. Philosophical arguments have their place in all this, but their role should not be overestimated. They are means to understanding, not the substrate of philosophy itself.

Antonaccio (2012, pp. 136–137) quite observantly connects Murdoch's philosophical project to that of Michel Foucault, but what she mainly has in mind is how both in different but interestingly related ways seek a philosophy that makes a difference for how we live and how we see things. This is a large and difficult theme, and I would be content to establish a humbler connection: the idea of philosophizing as an exercise. "But then," writes Foucault,

> …what is philosophy today—philosophical activity, I mean—if it is not the critical work that thought brings to bear on itself? In what does it consist if not in the endeavour to know how and to what extent it might be possible to think differently, instead of legitimating what is already known? … The 'essay'—which should be understood as the assay or test by which, in the game of truth, one undergoes change, and not a the simplistic appropriation of others for the purpose of communication—is the living substance of philosophy, at least if we assume that philosophy is still what it was in times past, i.e., an 'ascesis,' askesis, an exercise of oneself in the activity of thought. (Foucault 1990, pp. 8–9)

Murdoch's invitation in MGM is to an "*exercise of oneself in the activity of thought*". Stephen Mulhall talks about the book (borrowing a formulation form Murdoch herself) as making a great big "hall of reflection" (Mulhall 1997). This is true to the texts, in the sense that it accounts for the many byways and detours in the book, but it is also far too vague to characterize what she does in this specific chapter on consciousness in MGM. Murdoch here "thinks differently" and relocates the philosophical thinker from the magic circle of the western tradition (with its special concerns about knowledge, certainty, truth), to the seeking of selfless clarity in the mess of everyday living in a world of others (with suffering, longings, joys and also a peculiar void where God used to be). From this new location the question of truth, but also the question of goodness comes forth in a very different light.

12.4 Two Contexts for Platonic Ascendance

The Platonic image of ascendance from the Cave is used by Murdoch to characterize both the moral life and the pursuit of philosophy. The myth of the Cave provides a kind of universal imagery for the human person's quest for understanding or for a truthful vision.

> The Cave is a religious Myth suggesting, what is also accessible to any careful not necessarily philosophical reflection, that there are discernible levels and qualities of *awareness* and *experience* (we need this terminology), which cannot be reduced to acquaintance with neutral factual propositions or analysed in terms of dispositions to act. Of course, there are neutral scientific or scholarly or legal disciplines and procedures and states of mind, and these, often to be thought of as ideal limits, are essential and without them we would indeed 'perish and go to ruin'. But they represent one aspect only of the idea of truth and occupy a smaller area than is sometimes suggested by those who conjure up a vast world of facts in contrast to a small specialized activity of evaluating. Beside the idea of truth as some sort of mechanical accuracy (science is not really like this anyhow) or obvious, and of course necessary, daily reportage (the cat is on the mat), we need a larger idea which can contain, turning toward the individual, ideas of 'truthfulness' and 'wisdom'. (MGM, p. 183)

To overcome the world of shadows one must somehow be transformed, and this involves value and one's whole way of seeing things. The myth points to the ordinary though not necessarily easily attainable knowledge, that there are different qualities of awareness. It does not concern itself mainly with plain facts (though we need those too), but with the imaginative capacity to perceive differently, to grow as a human being, to become more truthful in one's observations. Murdoch uses the word "wisdom" with a casual confidence, defying the fact that it sounded dreadfully old-fashioned in late twentieth century moral philosophy. In the platonic story, truthfulness and goodness merge.

But it is, none the less, vital that passages like these are not read in terms of an overarching claim that morality and philosophy are one, that the ascent involved in these are one and the same, that philosophical self-transformation and progress means moral self-transformation and progress or vice versa. Although philosophy for Murdoch is work on oneself, a work on "how one sees things", on how one is conditioned by one's circumstances, prejudices, desires, she retains a place for philosophy as fairly technical pursuit, conducted within a tradition, answering questions which derive a certain part of their meaningfulness from within the tradition. When comparing philosophy with literature she notes that: "Philosophy aims to clarify and to explain, it states and tries to solve very difficult highly technical problems and the writing must be subservient to this aim. .." (Murdoch 1997, p. 4) She also emphasizes the importance of moral philosophy as a kind of specialized intellectual discipline.

> I also feel sure that moral philosophy ought to be defended and kept in existence as a pure activity, or a fertile area, analogous in importance to un-applied mathematics or pure 'useless' historical research. Ethical theory has affected society, and has reached as far as to the ordinary man, in the past, and there is no good reason to think that it cannot do so in the future. For both the collective and the individual salvation of the human race, art is doubtless more important than philosophy, and literature most important of all. But there can be no substitute for pure, disciplined, professional speculation: and it is from these two areas, art and ethics, that we must hope to generate concepts worthy, and also able, to guide and check the increasing power of science." (Murdoch 2001, p. 74)

Philosophical thought, when pursued in the ascetic spirit, as an exercise of oneself, means thinking very hard about what various philosophical and theoretical ideas really mean, against the background of our everyday lives and our cultural setting. It means that philosophical discussions cannot be taken at face value, as set tables, as board games waiting for the next move. The philosopher, as an individual thinker, has to take responsibility not only for his moves, but for the setting as a whole, the perspective it entails. And she also needs to move on, ascend from the cave, to seek a better framework, a truer setting when this is called for. This may require character strengths not irrelevant to morality. But philosophical and theoretical progress do not translate to moral virtue in our dealings with other people. Philosophers and thinkers can be quite extraordinarily apt in their capacity to push the boundaries of their thinking, to ascend from their theoretical caves, and yet remain obnoxious human beings.

A good person, again, can in Murdoch's view very well be an ordinary, non-intellectual person, quite unaffected by intellectual pursuits. Moral virtue lies in the unflinching attentiveness to other people and the world that one shares with them. For the intellectual person or the philosopher, this is something that requires arduous work in its own right, quite distinct from the work put into philosophy, and for this work the platonic myth of the cave may suggest helpful imagery. But Murdoch emphasizes that some people may have a special talent for attentiveness, and that this is not dependent on articulacy: this is illustrated in the characters of the selfless aunts and virtuous peasants that are part of our literature and folklore.

> 'Becoming better' is a process involving an exercise and refinement of moral vocabulary and sensibility. Yet we must also in discussing virtue, as distinct from practicing it, beware of seeming to suggest that the articulate educated man is better than the inarticulate uneducated man because he can think rationally and formulate and verbalise his distinctions. …Virtue shows in actions, goodness can be simple. Here the idea of imaginative grasp of one's surroundings may be preferred to that of a rational survey or ability to learn, or we may like to insist that good reasoning and learning is imaginative. The virtuous peasant can imagine the results of what he does and knows in his experience what truthfulness is. (MGM, p. 324)

This emphasis on simple, non-intellectual perceptiveness and natural virtue leans partly on observation: such people exist. Observational statements of this kind are quite typical for Murdoch's thinking. But this emphasis also plays a central role in the necessary self-criticism of the intellectual seeker, trying to make her way out of the cave. The idea of a necessary connection between intellectual progress and moral virtue is one of the most central fantasies that lure the intellectual, articulate person into a distorted view of his own moral standing. This is not to deny the value of education, intellectual capacity and articulacy, just to point out a treacherous path of reasoning. Many people, who lack the means or temperament for developing articulacy and theoretical reflexivity in moral matters, have a spontaneous capacity for moral sensitivity, generosity, warmth and discernment. The "imaginative grasp of one's surroundings" needed for goodness may be the fruit of innate talent or acquired in some way quite distinct from explicit reasoning.

12.5 Formulating an Intelligent Philosophy of Wisdom

This separation of moral improvement, from the transformative potential of philosophy and knowledge seeking, is important for Murdoch because it counteracts the potential self-satisfaction of philosophy of wisdom. About Plato she notes that "(t)he dialogues are full of confused truth seekers" (MGM, p. 228). The intellectual and philosopher, making sometimes rapid progress in their understanding of their subject matter, may be misled to think that moral progress ensues from this understanding. Conversely, a philosopher very sure of the moral righteousness and sincerity of his political or moral quest or standing may confuse this with quality in the contents of his philosophical or theoretical thinking.

Complacency of this kind is a common vice and not something that goes with a specific philosophical outlook. Yet it seems to be more easily accommodated in a perspective on philosophy as the pursuit of wisdom, where moral progress and theoretical, intellectual progress are seen as aspects of a unitary process.

Philosophies of wisdom are currently, once more, in fashion: In popular culture, in post-Wittgensteinian philosophy (philosophy as therapy), in feminist philosophy and in the broad interest in the philosophical schools of Hellenism and late antiquity, contemporary scholars and ordinary people seek ways of looking which are more holistic and more alive to the complexities of our lives than analytic philosophy tends to be. In this movement, we can see a fruitful desire to bring back philosophical inquiry to questions that seem to matter to us as ordinary thinking human beings. But what is central to the western tradition, a kind of disinterested pursuit of knowledge, the exercise of thought for its own sake, is easily either distorted or lost in a contemporary philosophy of wisdom. Thus, one challenge today, for the broad turn to philosophies of wisdom, is to preserve within their auspices what is worthy in this tradition. Murdoch, with her double use of the Platonic cave myth, is one of the philosophers who may help us articulate a reconciliation between philosophy as a quest of self-improvement and moral growth on the on hand, and as an (in some sense) impersonal pursuit of self-transformation in the realm of knowledge on the other hand. Keenly aware of both the fundamental intertwinement of knowledge and value in consciousness and the risks of indiscriminate merging of knowledge and goodness, she offers us a body of work where the intelligent balancing of these considerations is a constant concern.

References

Antonaccio, M. (2000). *Picturing the human: The moral thought of iris Murdoch*. Oxford: Oxford University Press.
Antonaccio, M. (2012). *A philosophy to live by: Engaging Iris Murdoch*. Oxford: Oxford University Press.
Broackes, J. (2011). Introduction. In J. Broackes (Ed.), *Iris Murdoch, philosopher*. Oxford: Oxford University Press.
Clarke, B. (2011). The prospect for critical moral perception. In J. Broackes (Ed.), *Iris Murdoch, philosopher*. Oxford: Oxford University Press.
Diamond, C. (1996). We are perpetually moralists: Iris Murdoch, fact, and value. In M. Antonaccio & W. Schweiker (Eds.), *Iris Murdoch: The search for human goodness*. Chicago: University of Chicago Press.
Forsberg, N. (2013). *Language lost and found*. New York: Bloomsbury.
Foucault, M. (1990). *The history of sexuality vol. 2: The use of pleasure*. New York: Vintage.
Gaita, R. (2003). *The philosopher's dog*. London: Routledge.
Hämäläinen, N. (2015). Reduce ourselves to zero? Murdoch, Lovibond and Feminism. *Hypatia, 30,* 743–759.
Lovibond, S. (2011). *Iris Murdoch, gender and philosophy*. London: Routledge.
Mulhall, S. (1997). Constructing a hall of reflection: Perfectionist edification in Iris Murdoch's "Metaphysics as a Guide to Morals". *Philosophy, 72,* 219–239.
Murdoch, I. (1997). *Existentialists and mystics—Writings in philosophy and literature*. London: Chatto & Windus.
Murdoch, I. (2001). *The sovereignty of good over other concepts*. London: Routledge.
Murdoch, I. (1992). *Metaphysics as a Guide to Morals*. London: Vintage.
Widdows, H. (2005). *Murdoch's morality*. Farnham: Ashgate.

Part IV
Celebrating Women Philosophers in Art

Chapter 13
Celebrating Women Thinkers

Sigridur Thorgeirsdottir

Catrine Val is an artist who has in the past decade studied women philosophers of the past and the present. She has travelled to Asia, Africa and South America to explore traditions of women's wisdom. Her work coincides with the present growing research into women philosophers. She was invited to present her pictures at the Helsinki conference on *Women in the History of Philosophy: Methodological Reflections* in 2015 where the main focus was on her works on women in the Western philosophical tradition.

In her works, Catrine Val celebrates the powerful presence of women thinkers throughout the ages. She reimagines their reappearance today with her artistic interpretations of them that are based on her research into their ideas and their times. Her work displays concretely how artistic and scientific research can complement each other as well as shed light on each other. Her artistic presentation of different thinkers and wisdom figures shows what she captures as an essential aspect of their philosophy, and how she translates it into modern imagery and styles of comportment, clothing and props to complete the picture. With her works, Catrine Val counters tendencies that she views as paradigmatic of contemporary popular culture. We live in aesthetic times in which we realize our individuality through self-expression, almost as if we create ourselves as our own works of art. She claims that our times are afflicted by a flood of narcissism, and an obsessive cult of self-expression. Our efforts of self-expression hence have an ambiguity to them. They may make us all similar, undermining strong individuality because narcissism disconnects ourselves from ourselves by mirroring ourselves primarily in others. Self-expression that is not narcissistic can, on the other hand, strengthen us in being ourselves. It is therefore not paradoxical that Catrine Val implements the imagery of the cult of self-expression, of staging of the self to portray women thinkers. She confronts narcissistic tendencies of self-expression with their own means for her explicit goal is to produce authentic

S. Thorgeirsdottir (✉)
Department of Philosophy, University of Iceland, 101 Reykjavik, Iceland
e-mail: sigrthor@hi.is

S. Thorgeirsdottir and R. E. Hagengruber (eds.), *Methodological Reflections on Women's Contribution and Influence in the History of Philosophy*, Women in the History of Philosophy and Sciences 3, https://doi.org/10.1007/978-3-030-44421-1_13

work in a world where everything is staged. There is therefore often an idiosyncractic feature to her figures, underscoring how these women thinkers were fiercely independent and true to themselves and ahead of their times in their outlook. Nevertheless, most of them became historical outsiders by not being admitted to the gallery of canonical figures in the history of philosophy. These philosophers have therefore not been properly acknowledged and appreciated for their contribution to philosophy. To address this imbalance Catrine Val offers an exclusively feminine perspective by presenting 'herself' through a series of images as the embodiment of various female philosophers, whose wisdom deserves recognition. In her own words:

> I present the philosophers in an iconographic way consistent with modern media discourse. This approach enables me to show how female philosophers have shaped their own reality and defined what it means to be a female philosopher. By exploring the visual similarities between past and current philosophers I reveal a timeless vision of the female philosopher. I present the images as realist fictions using transformations and replications to show the women in different contexts that cast light on the meaning and value of their ideas.

For further information on the works of Catrine Val, see http://www.catrineval.com/proj/philosopher-the-female-wisdom/.

Chapter 14
Catrine Val: Female Wisdom in Philosophy

Cao Wenyi

S. Thorgeirsdottir and R. E. Hagengruber (eds.), *Methodological Reflections on Women's Contribution and Influence in the History of Philosophy*, Women in the History of Philosophy and Sciences 3, https://doi.org/10.1007/978-3-030-44421-1_14

Emilie du Châtelet *(1706 - 1749)*

Anne Conway *(1631 – 1679)*

Diotima

Hannah Arendt (1906-1975)

Hypatia *(c. 350-370 - 415 AD)*

Iris Murdoch *(1919 - 1999)*

Christine de Pizan (1364 - 1429)

Simone de Beauvoir (1908 - 1986)

The manufacturer's authorised representative in the EU is Springer Nature Customer Service Centre GmbH, Europaplatz 3, 69115 Heidelberg, Germany. If you have any concerns regarding our products, please contact ProductSafety@springernature.com

Printed and bound by CPI Group (UK) Ltd, Croydon, CR0 4YY
05/10/2026
02228126-0002